Inoculating Cities, Volume II:
Case Studies of the Urban Response to the COVID-19 Pandemic

Inoculating Cities, Volume II:

Case Studies of the Urban Response to the COVID-19 Pandemic

Edited by

Rebecca Katz
Center for Global Health Science and Security, Georgetown University, Washington, DC, United States

Matthew R. Boyce
Center for Global Health Science and Security, Georgetown University, Washington, DC, United States

ELSEVIER

ACADEMIC PRESS
An imprint of Elsevier

Academic Press is an imprint of Elsevier
125 London Wall, London EC2Y 5AS, United Kingdom
525 B Street, Suite 1650, San Diego, CA 92101, United States
50 Hampshire Street, 5th Floor, Cambridge, MA 02139, United States
The Boulevard, Langford Lane, Kidlington, Oxford OX5 1GB, United Kingdom

Notices
Knowledge and best practice in this field are constantly changing. As new research and experience broaden our understanding, changes in research methods, professional practices, or medical treatment may become necessary.

Practitioners and researchers must always rely on their own experience and knowledge in evaluating and using any information, methods, compounds, or experiments described herein. In using such information or methods they should be mindful of their own safety and the safety of others, including parties for whom they have a professional responsibility.

To the fullest extent of the law, neither the Publisher nor the authors, contributors, or editors, assume any liability for any injury and/or damage to persons or property as a matter of products liability, negligence or otherwise, or from any use or operation of any methods, products, instructions, or ideas contained in the material herein.

ISBN: 978-0-443-18701-8

For Information on all Academic Press publications
visit our website at https://www.elsevier.com/books-and-journals

Publisher: Stacy Masucci
Acquisitions Editor: Elizabeth A. Brown
Editorial Project Manager: Michaela Realiza
Production Project Manager: Rashmi Manoharan
Cover Designer: Mark Rogers and Matthew R. Boyce

Typeset by MPS Limited, Chennai, India

Working together
to grow libraries in
developing countries

www.elsevier.com • www.bookaid.org

Contents

6. **Future-proofing: Quezon City's response to the COVID-19 pandemic as investments against the next epidemic**   115
    Noel Bernardo, Jaifred Christian Lopez, Esperanza Anita E. Arias and Misha Coleman

Section III
Workforce and surge capacity   141

7. **Expanding workforce surge capacity and the multijurisdictional response to the COVID-19 pandemic in the Las Vegas metropolitan area**   147
    Fermin Leguen, Cassius Lockett and Jeffrey Quinn

8. **Using mobile financial services to improve community
health workers' efficiency during the COVID-19
pandemic in Dhaka, Bangladesh**   167

*Farzana Misha, Syed Hassan Imtiaz, Margaret McConnell,
Richard Cash and Sabina Faiz Rashid*

Section IV
Vulnerable populations and pandemic response   187

9. **Prioritizing local context and expertise in a global
pandemic: the New Orleans response to COVID-19**   193

Jennifer Avegno, Kasha Bornstein and Jordan Vaughn

Section V
Risk communication

List of contributors

Adam Abadir Office of the Comptroller, State of Maryland, Annapolis, MD, United States

Akin Abayomi Division of Haematology, University of Stellenbosch, Western Cape Province, South Africa; Nigerian Institute of Medical Research, Nigeria Ministry of Health, Lagos State Secretariat, Lagos, Nigeria

Farouk F. Abou Hassan Medical Laboratory Sciences Program, Division of Health Professions, Faculty of Health Sciences, American University of Beirut, Beirut, Lebanon

Tunde Ajayi Nigerian Institute of Medical Research, Nigeria Ministry of Health, Lagos State Secretariat, Lagos, Nigeria

Munzer Alkhalil Research for Health Systems Strengthening Syria (R4HSSS), National Institute for Health Research, Centre for Conflict and Health Research, King's College London, London, United Kingdom

Esperanza Anita E. Arias Quezon City Health Department, Quezon City, National Capital Region, Philippines

Jennifer Avegno New Orleans Health Department, New Orleans, LA, United States; Department of Emergency Medicine, LSU Health Sciences Center New Orleans, New Orleans, LA, United States

Noel Bernardo International SOS, Pasig City, National Capital Region, Philippines; Atlantic Institute, Rhode Trust, Oxford, United Kingdom

Kasha Bornstein Department of Emergency Medicine, LSU Health Sciences Center New Orleans, New Orleans, LA, United States

Mirna Bou Hamdan Medical Laboratory Sciences Program, Division of Health Professions, Faculty of Health Sciences, American University of Beirut, Beirut, Lebanon

Gemma Bowsher Research for Health Systems Strengthening Syria (R4HSSS), National Institute for Health Research, Centre for Conflict and Health Research, King's College London, London, United Kingdom

Richard Cash Department of Global Health and Population, Harvard T.H. Chan School of Public Health, Harvard University, Cambridge, MA, United States

Bryan W.K. Chow Ministry of Health, Singapore, Singapore

Misha Coleman International SOS, Melbourne, VIC, Australia; Department of Population and Global Health, University of Melbourne, Melbourne, VIC, Australia

Letitia Dzirasa Baltimore City, Office of the Mayor, Baltimore, MD, United States

Abdulkarim Ekzayez Research for Health Systems Strengthening Syria (R4HSSS), National Institute for Health Research, Centre for Conflict and Health Research, King's College London, London, United Kingdom

Kimberly Eshleman Office of Public Health Preparedness and Response, Baltimore City Health Department, Baltimore, MD, United States

Marc Ho Ministry of Health, Singapore, Singapore; Saw Swee Hock School of Public Health, National University of Singapore, Singapore, Singapore

Syed Hassan Imtiaz BRAC James P Grant School of Public Health, BRAC University, Dhaka, Bangladesh

Parangimalai Diwakar Madan Kumar Department of Public Health Dentistry, Ragas Dental College and Hospital, Chennai, Tamil Nadu, India

Vernon Lee Ministry of Health, Singapore, Singapore; Saw Swee Hock School of Public Health, National University of Singapore, Singapore, Singapore

Fermin Leguen Southern Nevada Health District, Las Vegas, NV, United States

Cassius Lockett Southern Nevada Health District, Las Vegas, NV, United States

Jaifred Christian Lopez Department of Population Health Sciences, Duke University School of Medicine, Durham, NC, United States

Jennifer Martin Baltimore City Health Department, Baltimore, MD, United States

Margaret McConnell Department of Global Health and Population, Harvard T.H. Chan School of Public Health, Harvard University, Cambridge, MA, United States

Nada M. Melhem Medical Laboratory Sciences Program, Division of Health Professions, Faculty of Health Sciences, American University of Beirut, Beirut, Lebanon

Takako Misaki Kawasaki Institute for Public Health, Kanagawa, Japan

Farzana Misha BRAC James P Grant School of Public Health, BRAC University, Dhaka, Bangladesh

Nobuhiko Okabe Kawasaki Institute for Public Health, Kanagawa, Japan

Preeti Patel Research for Health Systems Strengthening Syria (R4HSSS), National Institute for Health Research, Centre for Conflict and Health Research, King's College London, London, United Kingdom

Saravanan Poorni Department of Conservative Dentistry and Endodontics, Sri Venkateswara Dental College and Hospital, Chennai, Tamil Nadu, India

Jeffrey Quinn Southern Nevada Health District, Las Vegas, NV, United States

Sabina Faiz Rashid BRAC James P Grant School of Public Health, BRAC University, Dhaka, Bangladesh

Tomoya Saito National Institute of Infectious Diseases, Tokyo, Japan

Zameer Shervani Food and Energy Security Research and Product Centre, Sendai, Japan

Patricia van der Ross Community Services and Health, City of Cape Town, South Africa

Jordan Vaughn Department of Emergency Medicine, LSU Health Sciences Center New Orleans, New Orleans, LA, United States

Wycliffe Wei Ministry of Health, Singapore, Singapore

Introduction

Matthew R. Boyce and Rebecca Katz
Center for Global Health Science and Security, Georgetown University, Washington, DC, United States

On December 27, 2019, the Hubei Provincial Hospital of Integrated Chinese and Western Medicine reported a series of pneumonia cases of unknown etiology or origin in the city of Wuhan—a megacity of over 10 million inhabitants located in Hubei Province, China [1]. The Wuhan city government subsequently arranged for public health and medical experts to conduct preliminary investigations into these cases—the results, of which, suggested that the disease was a viral pneumonia that may be linked to the Huanan Market in Wuhan. Three days later, on December 30, the Wuhan City Health Commission issued an urgent notice to local hospitals on the Treatment of Patients with Pneumonia of Unknown Cause; and on December 31 the National Government of China officially notified the World Health Organization (WHO) China country office of the evolving outbreak.

Over the next several weeks, the findings of the preliminary investigations were confirmed, and additional cases of the disease were identified and reported around the world. On January 7, 2020, the pneumonia was confirmed to be caused by a novel coronavirus, which was called 2019-nCoV (2019 novel coronavirus). Additional cases of the disease were confirmed and reported in travelers from Wuhan in Thailand (January 12, 2020), Japan (January 15, 2020), South Korea (January 20, 2020), and the United States (January 21, 2020).

On January 30, 2020, Dr. Tedros Adhanom Ghebreyesus, the director-general of the WHO, declared the outbreak a Public Health Emergency of International Concern [2]—a rare measure meant to signal to the world that an outbreak may constitute a global public health risk through the international spread of disease that could require an immediate and coordinated international response. By the end of the next day, 1 month after the WHO had been notified of the outbreak, there had been 9826 confirmed cases of the disease in 20 countries—a majority of which were located in China or had a travel history inclusive of the country.

On February 11, 2020, the WHO renamed the virus—changing the name from 2019-nCoV to SARS-CoV-2 (severe acute respiratory syndrome coronavirus 2)—and announced that the disease would be called COVID-19, which was an acronym for coronavirus disease 2019 [3]. And, 1 month later, on March 11, 2020, with over 118,000 cases and 4000 deaths reported across 114 countries, the WHO characterized the outbreak as a pandemic [4]. Two and a half years later, there were over 560 million confirmed cases of COVID-19 and 6.3 million associated deaths reported to the WHO, although these numbers likely represent underestimates of the true number of cases and the burden of the pandemic [5].

While the substantial risks—health, economic, social, and otherwise—posed by pandemics have long been recognized, the world was caught off guard by the COVID-19 pandemic. Based on previous experiences and risk assessments, much of the work in pandemic preparedness and response had focused on addressing the risks posed by another respiratory pathogen—influenza. Although most pandemic experts agreed that novel influenzas and novel coronaviruses were of deep concern, and most likely to cause the widespread morbidity and mortality.

Before the COVID-19 pandemic, there were six identified coronaviruses that caused illness in humans. Of these, four resulted in mild respiratory illness—similar to the common cold. The other two coronaviruses, however—SARS-CoV and MERS-CoV—cause more severe illness in humans. The pandemic potential of coronaviruses was first recognized with the emergence of a novel human coronavirus in 2002—SARS-CoV, which causes the disease known as SARS—that caused global alarm after some 8000 cases and nearly 800 deaths were reported in 2003 before the virus inexplicably disappeared from human populations [6].[1] While the origins of the outbreak remain elusive, it is believed to have started in southern China—likely in a market that sold both slaughtered and live animals for human consumption—before spreading to Hong Kong, Singapore, Vietnam, Canada, and onward [6].

Ten years later, in 2012, the coronavirus Middle East respiratory syndrome (MERS)-CoV was identified as the pathogen responsible for causing a severe, acute respiratory disease called MERS [7]. Since its initial discovery, there have been over 2400 laboratory-confirmed cases of MERS and over 850 deaths—the majority of which have occurred in the Kingdom of Saudi Arabia [8].[2] MERS is a zoonotic virus—meaning it is transmitted between animal and human populations—though there have been

---

1. Indeed, it was partially due to this outbreak that the International Health Regulations—a legally binding agreement of 196 countries to build the capacities required to detect, report, and respond to public health emergencies—were revised in 2005.

2. As of September 2019, Saudi Arabia has accounted for 2077 of the laboratory-confirmed cases and 773 of the related deaths—amounting to a case−fatality ratio of 37.2%.

documented instances of transmission between humans. The most notable instance of human-to-human transmission occurred in 2015 in the Republic of Korea when a man flew from the Middle East to Seoul—sparking an outbreak that lasted for several months and resulted in nearly 200 cases and 38 deaths, and what was considered at the time to be a large-scale quarantine of populations with school closures lasting weeks [9].

Still, there is more linking these outbreaks than the fact that they were all caused by related viruses—their spread, both nationally and internationally, occurred in cities and urban environments. The initial case of the 2002–03 SARS outbreak was retrospectively identified in November 2002 in the city of Foshan in Guangdong Province, China and had been detected in two additional cities in the province by mid-December [6]. It was subsequently detected in other major cities—Beijing, Hong Kong, Singapore, Taipei, and Toronto—that act as global hubs of travel and trade. Similarly, when MERS has been detected outside of the Middle East, it has most often been in individuals who recently traveled to the region and in cities—such as Florence, Italy; London, United Kingdom; and Monastir, Tunisia. And, when COVID-19 began to spread in early 2020, it was detected in other cities in China, before it was detected urban environments in other countries, including Bangkok, Thailand; Kanagawa Prefecture, Japan;[3] Gwangju and Seoul, South Korea; and Seattle, United States.

This is not surprising in itself. Given the role that cities play as transportation hubs in today's globalized world, one might expect a majority of imported cases to be detected and diagnosed in cities. However, there are other characteristics and considerations that render cities and urban environments especially prone to infectious disease outbreaks—which may ultimately render them the factor that determines whether an outbreak remains small and localized, or grow into larger epidemics or pandemics.

Indeed, this has been seen throughout the response to the COVID-19 pandemic. Early in the pandemic, local political authorities were often the entities primarily responsible for determining what actions were needed to respond to the outbreak. And, as time went on, they were frequently responsible for implementing guidance and directives from higher levels of government. These actions—intended to curb the spread of disease and mitigate the health impacts of the pandemic—were occasionally drastic and sometimes difficult for the public to comprehend. For example, before the pandemic, few could have likely imagined the aggressive measures that were taken by the Chinese government—which effectively included quarantining entire cities and an estimated 56 million people. Others were less extreme and included things like holding daily press conferences to update the populations about the local situation. Regardless, the experiences in responding to

---

3. An urban, coastal prefecture that borders Tokyo to the south

the COVID-19 pandemic have led to a renewed interest in the roles and authorities of subnational actors in responding to infectious disease outbreaks and public health emergencies.

Cities and the response to infectious disease events: from localized outbreaks to pandemics

Before discussing the roles of cities and local officials in the response to infectious disease outbreaks, it is important to differentiate and clarify terms that are frequently used to describe them—epidemics and pandemics. The word epidemic is derived from two Greek words—*epi* (meaning "upon") and *demos* (meaning "people")—and refers to the occurrence of an illness, condition, health-related behavior, or other health-related events in excess of what would normally be expected in a specific area or population over a specific time. The term outbreak is frequently used to refer to the same phenomenon—and functionally there is no difference between an epidemic and an outbreak—but it is sometimes used to avoid the connotations associated with the word epidemic that may evoke public panic or to communicate that an epidemic is more localized in nature.

Similarly, the word pandemic is derived from the Greek words *pan* (meaning "all") and *demos* (meaning "people") and, broadly speaking, refers to an epidemic occurring over a large geographic area. While there is no official requirement of the geographic scope required to characterize an outbreak as a pandemic, it is commonly understood to mean that an outbreak is affecting a large number of people and occurring across multiple countries and continents.[4] Notably absent from the definition of a pandemic, however, is an assessment of disease severity or the risk posed by an outbreak. Put another way, the key difference that distinguishes a pandemic from an epidemic is not in the severity or risk, but in the degree to which an outbreak has spread geographically.[5] Still, at the time of writing, this very definition is being reexamined by global health officials, as the community assesses development of appropriate triggers for action in a strengthened global health security architecture.

The role of local authorities in public health and the response to infectious disease outbreaks are widely recognized but vary according to the

---

4. According to this definition, infectious disease outbreaks such as Ebola in 2014 and Zika in 2015 are epidemics because while they affected a large number of people and several countries, the countries were all located in the same region and continent; this compared to COVID-19, which has affected a large number of people in several countries and continents.
5. The declaration of a Public Health Emergency of International Concern, on the other hand, does represent a risk assessment. Thus, while there was much media attention given to the WHO's characterization of COVID-19 as a pandemic on March 11, 2020, the real alarms should have been sounded when the WHO characterized the outbreak as a Public Health Emergency of International Concern on 30 January 2020.

political context in which the city exists and operates. This means that there is a substantial amount of heterogeneity in the specific responsibilities and authorities of local officials and authorities globally, but also occasionally within the same country. However, broadly speaking, local authorities maintain some role in activities, including assessing and monitoring population health and risks, investigating and responding to emerging health problems and hazards, building and maintaining workforces, partnering with communities and local organizations to improve health, communicating with local populations, and working to provide equitable access to health and health services. All of these considerations are essential for responding to infectious disease outbreaks, irrespective of the scope and size.

Nonetheless, as outbreaks grow, additional actors and authorities are likely to be involved. For smaller, localized events, the local authorities may well be the only entities involved in a response. But when localized outbreaks develop into larger epidemics that impact broader geographic areas, other subnational authorities and even national, regional, or international authorities may be involved in the response. This can complicate response efforts as authorities and actors may have overlapping mandates that make specific responsibilities murky. And, if a pathogen causes severe disease, national authorities will almost certainly be involved, particularly as the resulting economic and societal consequences could render the outbreak a matter of national security.

The structure of this book

This book was born out of a call to capture the experiences of municipalities around the world in responding to the COVID-19 pandemic and to document several of the innovative ways in which cities were responding to the pandemic. As referenced previously, while the risks posed by pandemics, and infectious diseases more broadly, had long been recognized, the emergence of the SARS-CoV-2 virus and resulting pandemic caught the world off guard. This resulted in authorities at all levels scrambling to mount pandemic responses, which were all too often marked with indecision and confusion. Accordingly, there was a clear need to develop an evidence base informed by case studies from this response to help prepare the world for future epidemics and pandemics.

We launched a broad call for chapters in 2021 and invited researchers and local authorities to reflect on their efforts to respond to the COVID-19 pandemic in the cities they call home. We welcomed chapters that dealt with a wide variety of topics, as well as those that featured perspectives and voices not typically featured in scholarly works. The result is a rich collection of chapters, written by a diverse group of authors, containing accessible information for academics, policymakers, and professionals alike.

These chapters are broadly organized into five themes: pandemic governance and coordination, technology and digital approaches, workforce and surge capacity, vulnerable populations, and risk communication. It is important to note, however, that these themes are somewhat arbitrary. Indeed, as will soon become clear to readers, many of the chapters included in this volume discuss multiple themes. A chapter included in the workforce and surge capacity section may discuss how technology was used, while chapters in the governance section may discuss vulnerable populations. Given the scale of the pandemic and the whole-of-society approach that the response demanded, this is unsurprising.

Fittingly, the first chapter on pandemic governance and coordination describes the experiences of Kawasaki City, Japan. As detailed by Misaki and colleagues, in February 2020, Kawasaki City played a role in responding to one of the first large clusters of COVID-19 when authorities helped to support the response to an outbreak onboard the Diamond Princess cruise ship. The chapter then goes on to discuss the city's experiences in developing diagnostic capacities, preparing for vaccination campaigns, and responding to the first case of the Omicron variant.

The second chapter in this section discusses the pandemic response in Lagos, Nigeria. Abayomi and Ajayi discuss how the experiences from the 2014 Ebola epidemic resulted new emergency preparedness and biosecurity policies that provided a foundation for the response to the COVID-19 pandemic. More specifically, because of these efforts, Lagos was able to rapidly respond to the outbreak by utilizing its centralized incident command structure and the emergency operations center.

In the following chapter, Melhem and colleagues discuss the pandemic response in Beirut, Lebanon and how the response was complicated by the need to simultaneously respond to another public health emergency—the Beirut Port explosion in August 2020. These experiences have revealed a need to reimagine preparedness in the country and to create a national strategy that cuts across emergency preparedness, response, and recovery from crises, while giving due consideration to subnational implementation.

The final chapter in the governance section details the governance of the pandemic response in Idlib, Syria. In this chapter, Ekzayez and colleagues discuss how the city—the last stronghold for opposition forces engaged in conflict with the Government of Syria—adopted a "delayed approach" in a context defined by scant resources. The authorities in the city also collaborated extensively with a variety of organizations, including the WHO and the White Helmets, to bolster the local response.

The technology and digital approaches section begins with a chapter by Wei and colleagues on the response in Singapore. The authors detail how a variety of the aspects of pandemic response, but most notably contact tracing, quarantine, and vaccination, were improved using various technologies and digital approaches including mobile applications and web-based platforms.

This chapter is followed by one in which Bernardo and colleagues discuss the response to the pandemic in Quezon City, Philippines. This particular response—one backed by strong local leadership and robust public–private partnerships—drew heavily on technology to implement an innovative response to the public health emergency. Importantly, the authors also highlight how these efforts have bolstered local resilience against future infectious disease outbreaks.

In the next chapter, Leguen, Lockett, and Quinn begin the workforce and surge capacity section by discussing how Las Vegas, United States surged the health-care workforce and other critical response capacities. Recognizing that COVID-19 surge capacity planning required active participation by diverse community sectors, public–private partnerships between local, state, tribal, and federal agencies were leveraged to enhance communications, improve operational coordination, and bolster access to critical public health resources and services.

This section concludes with a chapter by Misha and colleagues on how community health workers were used to support the pandemic response in Dhaka, Bangladesh. Briefly, these individuals were used to provide routine and pandemic-centered health-care services, raise awareness surrounding COVID-19, distribute facemasks, and support vaccination campaigns in hard-to-reach urban populations. The use of mobile financial services also factored heavily into these efforts, simultaneously working to improve the efficiency of operations and support the livelihoods of community health workers.

The next section begins with a chapter by Avegno, Bornstein, and Vaughn, discussing how the response in New Orleans prioritized the local context to guide response efforts and to direct resources as a means of promoting an equitable response that met the needs of some of the city's most vulnerable residents. As detailed by the authors, because of historical legacies, certain populations in the city have been systemically disenfranchised and concerted efforts were made by local officials to ensure that these individuals had access to testing services, and later vaccines.

This chapter is followed by van der Ross, in which she describes the experiences from Cape Town, South Africa. This chapter emphasized both the confusion surrounding the pandemic response, as well as the devastating impact that certain response measures had on vulnerable populations. Additionally, while there was a clear and demonstrated need to formulate detailed strategies for the health considerations of the pandemic response, there was also a need to develop strategies for healing the broader, societal traumas that were associated with response efforts.

The next chapter features a discussion by Kumar, Poorni, and Shervani on how the pandemic response was implemented in one of the largest slums in the world—the Dharavi Slum in Mumbai, India. This chapter details how many of the recommended response measures, such as social distancing, were impossible to implement in the densely populated and resource-limited urban environment. However, the government recognized the importance and moral imperative of including the slum population in the response efforts and used a proactive approach to target this population—one that included massive testing and contact tracing campaigns, as well as efforts to facilitate compliance with quarantine and isolation measures.

The final chapter of the book discusses pandemic risk communication efforts in Baltimore, United States. In this chapter, Dzirasa and colleagues provide an overview of several risk communication campaigns implemented in the city, including those that sought to increase access to diagnostic testing, improve vaccine uptake, and address mis- and disinformation. Further, and of importance, these efforts were tailored to local communities by emphasizing transparency and trust.

As we have seen over the course of the past several years, the pandemic was an incredibly dynamic situation. Nowhere in the world was left untouched by the pandemic and no city implemented a perfect response. Areas that were lauded for their responses early on faltered in responding to subsequent waves of infection, and areas that were initially devastated by the pandemic responded by doubling down on efforts to implement successful responses. As such, the timing of this volume is of importance. We have encouraged authors to ground their reflections as a means of providing context, but also wish to acknowledge the reality that much may have changed when this volume is available on physical and virtual bookshelves.

Still, we hope this volume ultimately contributes to the growing evidence base for best practices in response to infectious disease events. Not all responses are appropriate for all events, and each of the cities described in this volume is unique. But we believe the experiences described herein, and the examples of response to the same pathogen in cities around the world will be a critical resource for sharing lessons and improving epidemic and pandemic preparedness and response moving forward.

References

[1] Fighting COVID-19: China in action: The State Council Information Office of the People's Republic of China, <http://www.scio.gov.cn/zfbps/32832/Document/1681809/1681809. htm>; 2020 [accessed 04.05.22].

[2] Ghebreyesus T. WHO Director-General's statement on IHR Emergency Committee on Novel Coronavirus (2019-nCoV). Geneva: World Health Organization; 2020.

[3] World Health Organization. Naming the coronavirus disease (COVID-19) and the virus that causes it, <https://www.who.int/emergencies/diseases/novel-coronavirus-2019/technical-guidance/naming-the-coronavirus-disease-(covid-2019)-and-the-virus-that-causes-it>; 2020 [accessed 04.05.22].

[4] Ghebreyesus T. WHO Director-General's opening remarks at the media briefing on COVID-19 - 11 March 2020. Geneva: World Health Organization; 2020.

[5] World Health Organization. WHO Coronavirus (COVID-19) Dashboard, <https://covid19.who.int/>; 2020 [accessed 04.05.22].

[6] Institute of Medicine (US) Forum on Microbial Threats. In: Knobler S, Mahmoud A, Lemon S, Mack A, Sivitz L, Oberholtzer K, editors. Learning from SARS: preparing for the next disease outbreak. Washington (DC): National Academies Press (US); 2004.

[7] Zaki AM, van Boheemen S, Bestebroer TM, Osterhaus AD, Fouchier RA. Isolation of a novel coronavirus from a man with pneumonia in Saudi Arabia. New England Journal of Medicine 2012;367(19):1814–20. Available from: https://doi.org/10.1056/NEJMoa1211721.

[8] World Health Organization – Regional Office for the Eastern Mediterranean. MERS Situation Update, September 2019. Cairo: World Health Organization – Regional Office for the Eastern Mediterranean; 2019.

[9] Oh MD, Park WB, Park SW, Choe PG, Bang JH, Song KH, et al. Middle East respiratory syndrome: what we learned from the 2015 outbreak in the Republic of Korea. Korean Journal of Internal Medicine 2018;33(2):233–46. Available from: https://doi.org/10.3904/kjim.2018.031.

Section I

Pandemic governance and coordination

What is governance?

Governance describes how a governing authority coordinates policy implementation with other government institutions, the private sector, and broader society [1,2]. Health governance encompasses both efforts to strengthen health systems and coordination during an outbreak response. Because nonpharmaceutical interventions and illness affected nearly every aspect of daily life, effective health governance required a whole-of-government and a whole-of-society approach [3]. Organizations and ministries responsible for socioeconomic relief, education, and other affected fields were often included in pandemic response efforts because they could help implement nonpharmaceutical interventions and alleviate various societal stressors.

What is good governance?

Good governance during the pandemic response required broad multisectoral coordination. Policymakers integrated advice from scientists, peer-reviewed literature, civil society organizations, and nonhealth sectors that implemented relief measures and regulations (e.g., finance, education, transportation) [4]. Including the perspective of specialists in fields indirectly affected by an outbreak was particularly important for addressing the effects of social isolation, chronic conditions resulting from the disease, and other consequences of the disease.

For instance, various tools and checklists have recommended the creation of multisectoral mechanisms and task forces with representatives from various groups to improve coordination during the response to outbreaks [5,6]. As envisioned by the World Health Organization, a multisectoral task force could include a centralized authority that would work to ensure consistency between risk communication and mitigation policies (e.g., quarantine, vaccination requirements), manage resource distribution, and provide social services where needed [6]. This type of coordination was also necessary at the subnational level; urban health leaders require the authority to adapt response measures to the unique conditions they face. Importantly, it was advised that the authority strives for participation (i.e., include mechanisms for feedback from civil society) and transparency in its decision-making to maintain legitimacy and achieve effective governance [7].

Why does governance matter?

Leading global health security indices failed to accurately predict which countries were best prepared to respond to a disease outbreak (e.g., the World Health Organization's Joint External Evaluation Tool) [8,9]. Weak leadership and conflict between national and subnational entities stifled response efforts in otherwise prepared countries, resulting in remarkably high per capita COVID-19 case totals [9]. Leaders of countries that have mounted poor responses, relative to similar, peer countries, questioned the existence of the virus, discouraged compliance with nonpharmaceutical interventions, fostered disinformation about vaccinations and treatment, and even failed to abide by their own regulations [10–13]. The resulting disparities in purported pandemic preparedness and response capacities and COVID-19 responses have illustrated the necessity of good governance in an effective outbreak response.

The importance of governance for urban pandemic response

High population density increased a locality's potential to amplify infectious disease spread and quickly overwhelm health care capacities because nonpharmaceutical interventions like social distancing and isolation were made difficult. Moreover, vulnerable populations—especially those living in overcrowded slums—often needed to congregate around clean water sources, face poor sanitization conditions, and did not have the space to social distance, let alone isolate when infected [14].[1]

Lockdowns, quarantine/isolation, social distancing, and other nonpharmaceutical interventions are only effective with collective action. Effective governance that integrated both scientific evidence and feedback from affected populations was vital to implementing reasonable policies that effectively mitigate disease spread in urban environments.

Chapter introductions

The following chapters outline how regional health officials coordinated an effective response in a diverse set of contexts: Kawasaki City, Japan; Lagos, Nigeria; Beirut, Lebanon; and Idlib, Syria.

---

1. See also Chapter 11 of this book.

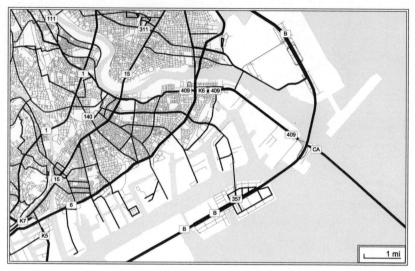

FIGURE 1   A map of Kawasaki City, Kanagawa Prefecture, Japan. Base map sourced from OpenStreetMap, using materials made available under CC BY-SA 2.0.

Kawasaki City is a government-designated city of over 1.5 million people in the Greater Tokyo area (Fig. 1). This chapter provides a brief summary of the public health system in the city before describing the response to the COVID-19 pandemic—highlighting the responses to early clusters of COVID-19, the development of laboratory diagnostic capacities, mass vaccination exercises, and an intensive response to Japan's first domestic case of the Omicron variant.

Lagos—a Nigerian mega-city with a rapidly growing population—was the epicenter of the COVID-19 outbreak in Nigeria (Fig. 2). This chapter discusses how previous experiences with epidemic response improved the governance of public health emergency response. More specifically, following the Ebola epidemic in 2014, the Lagos State Government created a policy on emergency preparedness and biosecurity that laid a foundation for overseeing and coordinating emergency preparedness plans. By educating key personnel, creating a reliable surveillance system, and establishing a biosafety level 3 laboratory and biobank, emergency response capabilities were increased. As a result of these actions, when the COVID-19 pandemic began, Lagos was able to quickly respond to the outbreak by utilizing the centralized incident command structure and the primary functions of the emergency operations center. Additional partnerships with the private sector, community involvement, and political commitment all helped to make the outbreak response effective.

Chapter three focuses on the governance of the pandemic response in Beirut, the capital city of Lebanon (Fig. 3). This chapter discusses how

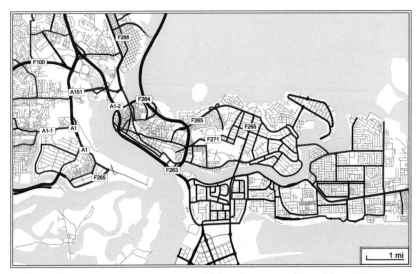

FIGURE 2 A map of Lagos, Nigeria. Base map sourced from OpenStreetMap, using materials made available under CC BY-SA 2.0.

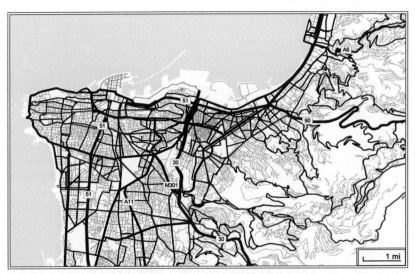

FIGURE 3 A map of Beirut, Lebanon. Base map sourced from OpenStreetMap, using materials made available under CC BY-SA 2.0.

pandemic response efforts were complicated by deteriorating socioeconomic conditions in the country, as well as another public health emergency—a large explosion in the Port of Beirut that resulted in a significant number of causalities and critical health care infrastructure. This chapter explains the evolution of the COVID-19 pandemic in Lebanon, paying

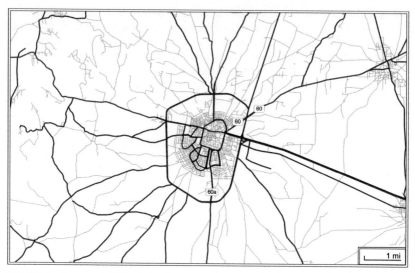

FIGURE 4 A map of Idlib, Syria. Base map sourced from OpenStreetMap, using materials made available under CC BY-SA 2.0.

special attention to efforts in Beirut, and discusses the gaps and challenges pertinent to pandemic preparedness and response.

The final chapter in this section discusses the pandemic response in the city of Idlib, Syria (Fig. 4). Idlib is the capital of the Idlib Governorate in Northwest Syria and represents the last stronghold for opposition forces engaged in over a decade of conflict with the Assad regime. The city has been subjected to the horrors of war, while huge numbers of people have been forcibly displaced within the region. This chapter discusses how the autonomous health authority in the city prepared for the arrival of the disease by adopting a "delay approach." Border closures and public awareness characterized early responses to the impending pandemic threat, and volunteer-led testing and contact tracing programs were scaled up in collaboration with the WHO-led health cluster based in Turkey.

References

[1] Assefa Y, Woldeyohannes S, Cullerton K, Gilks CF, Reid S, Van Damme W. Attributes of national governance for an effective response to public health emergencies: lessons from the response to the COVID-19 pandemic. Journal of Global Health 2022;12:05021. Available from: https://doi.org/10.7189/jogh.12.05021.

[2] Undesa, UNDP, UNESCO. Governance and Development: UN System Task Team on the Post-2015 UN Development Agenda. Geneva: United Nations; 2012.

[3] Kickbusch I, Gleicher D. Governance for health in the 21st century. Copenhagen: World Health Organization, Regional Office for Europe; 2013.

[4] Rajan D, Koch K, Rohrer K, Bajnoczki C, Socha A, Voss M, et al. Governance of the Covid-19 response: a call for more inclusive and transparent decision-making. BMJ Global Health 2020;5(5):e002655. Available from: https://doi.org/10.1136/bmjgh-2020-002655.

[5] Boyce MR, Katz R. Rapid urban health security assessment tool: a new resource for evaluating local-level public health preparedness. BMJ Global Health 2020;5:e002606. Available from: https://doi.org/10.1136/bmjgh-2020-002606.

[6] World Health Organization. Practical actions in cities to strengthen preparedness for the COVID-19 pandemic and interim checklist for local authorities. Geneva: World Health Organization; 2020.

[7] Fidler DP. Germs, governance, and global public health in the wake of SARS. Journal of Clinical Investigation 2004;113(6):799−804. Available from: https://doi.org/10.1172/JCI21328.

[8] Haider N, Yavlinsky A, Chang YM, Hasan MN, Benfield C, Osman AY, et al. The Global Health Security index and Joint External Evaluation score for health preparedness are not correlated with countries' COVID-19 detection response time and mortality outcome. Epidemiology and Infection 2020;148(e210):1−8. Available from: https://doi.org/10.1017/S0950268820002046.

[9] Baum F, Freeman T, Musolino C, Abramovitz M, De Ceukelaire W, Flavel J, et al. Explaining covid-19 performance: what factors might predict national responses. British Medical Journal 2021;372:n91. Available from: https://doi.org/10.1136/bmj.n91.

[10] Phillips T. "Police call for Bolsonaro to be charged for spreading Covid misinformation.". The Guardian Aug 2022;18.

[11] Paz C. All the president's lies about the coronavirus. The Atlantic 2020; 02 Nov.

[12] Astapenia R, Marin A. Belarusians left facing COVID-19 alone. London: Chatham House; 2020.

[13] Press TA. A new report blames Boris Johnson for allowing parties during COVID lockdown. NPR 2022; 25 May.

[14] Auerbach AM, Thachil T. How does Covid-19 affect urban slums? Evidence from settlement leaders in India. World Development 2021;140:105304. Available from: https://doi.org/10.1016/j.worlddev.2020.105304.

Chapter 1

COVID-19 control strategy and measures in Kawasaki City, Japan

Takako Misaki[1], Tomoya Saito[2] and Nobuhiko Okabe[1]
[1]Kawasaki Institute for Public Health, Kanagawa, Japan, [2]National Institute of Infectious Diseases, Tokyo, Japan

Background

Japan is an island country located in East Asia with an area of 378,000 km^2 and a population of 126 million. The country is divided into 47 prefectures, and each prefecture is further divided into municipalities and special districts. Among the municipalities in Japan, there are 20 that are ordinance-designated with a population of greater than 500,000 persons, and Kawasaki City is one of these cities.[1]

Kawasaki City contains seven districts and is a long and narrow city—measuring about 31 km east-west and about 19 km north-south—located in the northeast of Kanagawa Prefecture. The city is adjacent to Tokyo, the capital city of Japan, and near Yokohama, another large ordinance-designated city in the Kanagawa Prefecture. Kawasaki City is home to over 1.5 million people and witnesses the daily movements of many people, both domestically and abroad. The East Japan Railway passes through the city from north to south, and five other private railways also transect it. The city is home to Kawasaki Port in the southern part of the city and is very close to Tokyo International Airport (Haneda Airport), which is located across the Tama River. Approximately 3% of the residents in Kawasaki City are foreigners, which is slightly higher than the national average of 2.2% [1].[2]

---

1. In Japan, cities with a population of 200,000 or more are designated as core cities, while those with a population of 500,000 or more are designated by ordinances. At the time of writing, there are 62 core cities and 20 ordinance-designated cities.
2. The estimate for the percentage of foreign residents in Kawasaki City is from March 2021, while the national estimate is from October 2020.

This chapter focuses on several key measures implemented in Kawasaki City, Japan in response to the COVID-19 pandemic.

Public health and infectious disease control in Kawasaki City

In Japan, municipalities are the entities primarily responsible for the administration of public health activities for their local populations, including infectious disease control efforts. Each municipality or municipal district has an established public health center (PHC) that works with the national government to respond to infectious disease outbreaks. There are also local public health institutes (LPHIs) that exist, in part, to establish local infectious disease surveillance centers to collect and analyze infectious disease data. As of January 1, 2022, 84 LPHIs have been established in prefectures and designated cities throughout the country to conduct administrative tasks and examine specimens transferred from medical facilities through the PHCs. In cooperation with the National Institute of Infectious Diseases (NIID) and the Central Infectious Disease Surveillance Centers, the local infectious disease surveillance centers established by LPHIs conduct infectious disease surveillance and epidemiological surveys. In Kawasaki City, this public health system is implemented through the work of the Kawasaki City Institute for Public Health, the city's LPHI, the Kawasaki City Infectious Disease Surveillance Center, and seven district health center branches that work with the municipal PHC to support the city's public health system. Kawasaki City also houses three municipal hospitals—a university hospital that offers more than 1000 beds, and two regional hospitals with more than 700 beds.

The legal foundations for infectious disease control in Japan are outlined in the Act on the Prevention of Infectious Diseases and Medical Care for Patients with Infectious Diseases (the Infectious Diseases Control Law) [2]. The Infectious Diseases Control Law was promulgated in 1998 and entered into force in 1999 to ensure that measures for the prevention of and response to infectious diseases are harmonized and to give appropriate consideration to the human rights of patients. As outlined in Article 6 of the Infectious Diseases Control Law, diseases are classified into hierarchical categories numbered one through five. Category 1 Infectious Diseases include Crimean-Congo hemorrhagic fever, Ebola, Lassa fever, Marburg, plague, SARS, and smallpox; Category 2 Infectious Diseases include cholera, diphtheria, polio, shigella, and typhoid fever and paratyphoid fever; Category 3 Infectious Diseases include enterohemorrhagic *Escherichia coli* infections; Category 4 Infectious Diseases include known infectious diseases that are transmitted via animals, or carcasses thereof, food, drinks, clothing, bedclothes, or other personal property and affect the health of the people (i.e., hepatitis A, hepatitis E, highly pathogenic avian influenza, Q fever, rabies, and yellow fever); and Category 5 Infectious Diseases include other known infectious diseases such as HIV/AIDS, chlamydia, cryptosporidiosis, influenza (excluding highly pathogenic avian influenza), measles, drug-resistant staphylococcus infections, syphilis, and viral hepatitis (excluding hepatitis A and hepatitis E). The law also differentiates between "Designated Infectious Diseases," which are those recognized by the government as diseases that may threaten the lives of or seriously affect the health of people (excluding Category 1−3 Infectious Diseases), "Pandemic Influenza and New Infectious Diseases," and "New Diseases," which are those that demonstrated human-to-human transmission and hold the potential to cause an epidemic of that may threaten the lives of or seriously affect the health of the people. Under this law, physicians are required to report all cases of Category 1−4 Infectious Diseases and select Category 5 Infectious Diseases to the prefectural governor via the PHC, who then notifies the Minister of Health, Labour and Welfare.

In addition to these surveillance activities, there are designated medical institutions that implement sentinel surveillance to monitor the incidence of the target infectious diseases over designated periods of time (i.e., weeks or months). Six categories of sentinels are defined: pediatric sentinel sites, influenza sentinel sites, ophthalmology sentinel sites, sexually transmitted infection sentinel sites, designated infection sentinel sites, and suspected disease sentinel sites. These categories target many of the five categories of infectious diseases, as well as suspected disease carriers, as stipulated by the Ministry of Health, Labour and Welfare.

The 2012 Act on Special Measures for Pandemic Influenza and New Infectious Diseases Preparedness and Response is another important law relating to the control of infectious diseases. This Act was passed in the aftermath of the 2009 influenza pandemic and seeks to ensure the effectiveness of response measures and clarify the legal basis for various countermeasures [3]. Of importance, this law applies not only to new strains of influenza but also to outbreaks of emerging and re-emerging infectious diseases that can rapidly spread and have a serious impact on the public. Should an infectious disease rapidly spread across the country and have significant societal and economic impacts, the government can issue a "Declaration of Emergency Situation Regarding Pandemic Influenza and New Infectious Diseases," which allows for it to make requests to slow or halt the spread of disease that are implemented on a prefectural basis (e.g., that mass gathering events be suspended, the suspension of business operations in facilities where large numbers of people congregate, among others). If these requests are not complied with, the government may issue notices and fees to violators.

The response to COVID-19 in Kawasaki City

The causative agent of COVID-19 was identified as SARS-CoV-2 during the very early stage of the pandemic [4]. However, when the first case was detected and reported in Japan on January 16, 2020, COVID-19 had not yet been classified under the Infectious Disease Control Law. As a result, the responsibility for disease surveillance for this emerging disease had fallen upon reports from suspected disease sentinel sites that are mandated to report an undiagnosed serious infectious illness. It was on February 1, 2020, when COVID-19 was officially designated as a Designated Infectious Disease through a Cabinet Order, that rapid control and response measures (e.g., recommendations for hospitalization and restrictions on employment) could be implemented.

Japan and Kawasaki City gained experience in responding to COVID-19 relatively early in the outbreak by supporting the outbreak response to a disease cluster onboard the Diamond Princess cruise ship. This large cruise ship was brought to port in February 2020 in Yokohama Harbor, which is located in Tokyo Bay, adjacent to Kawasaki City. Some 3,700 passengers were on the vessel when it became clear that COVID-19 was spreading onboard, resulting in the isolation of private rooms onboard the vessel beginning on February 5, 2020. This prompted the mayor to establish a city-level advisory committee to inform the response to the pandemic, with the first meeting occurring that same day. Medical and disaster preparedness personnel traveled from all over Japan to support the response, but by February 19, there were as many as 619 confirmed cases onboard. While Yokohama City is a large city with a population of over 3.7 million people, no medical or public health system in the world maintains the capacity to accommodate all infected persons on a cruise ship. Accordingly, the ship was quarantined in the harbor, with severely ill persons disembarked and sent to various medical institutions and facilities across Japan, including at least 30 infected individuals being transported to medical institutions located in Kawasaki City. As a result, Kawasaki City was one of the first jurisdictions outside of China to address a large cluster of COVID-19 cases. Fortunately, there were no secondary or tertiary infections that resulted from this response, but by the time the Diamond Princess exited Yokohama Harbor on March 25, at least 712 of the passengers or crew had been infected and discharged from isolation. Because this outbreak occurred very early in the COVID-19 pandemic, and given the challenges associated with the location of the outbreak, preparations were difficult, and the response was one generally characterized by confusion. However, this experience also provided a learning opportunity for the medical institutions in Kawasaki City and a foundation for responding to the domestic waves of infection that would come.

Japan experienced three distinct waves of COVID-19 relatively early in the pandemic—with peaks in April 2020, August 2020, and January 2021—and the government recognized that the outbreak was likely to continue. As a result, in February 2021, because of the prolonged nature of the outbreak and the substantial efforts required to effectively coordinate and respond to the health emergency, the government reclassified COVID-19 from a Designated Infectious Disease to a Pandemic Influenza and New Infectious Disease through a revision of the Infectious Diseases Control Law. Reclassifying the disease allowed the government to more consistently and reliably promote pandemic response measures that it would not have been able to otherwise, as the responses to Designated Infectious Disease outbreaks are time-limited. This action proved providential, as the country subsequently experienced four additional waves of infection with peaks in May 2021, August 2021, February 2022, and August 2022. Each prefecture has subsequently taken measures to prevent the spread of COVID-19 by declaring a state of emergency and implementing priority response measures, such as requesting that restaurant industry businesses with a high risk of infection change their business hours and establishing fines for violating the order [5]. As a result, the pandemic response measures implemented in Kawasaki City for COVID-19 have been taken in close cooperation with Kanagawa Prefecture.

The responses to domestic clusters of COVID-19

Different lineages of SARS-CoV-2 have dominated the different epidemic waves in Japan. The first wave of the outbreak (April 2020) was characterized by the B.1.1 variant, which was imported from Europe. Later waves were dominated by B.1.1.214 variant (August 2020), the B.1.1.284 variant (January 2021), the B.1.1.7 alpha variant (May 2021), the B.1.617.2 delta variant (August 2021), the BA.1/BA.2 Omicron variants (February 2022), and the BA.5 Omicron subvariant (August 2022) [6]. Kawasaki City experienced these epidemic waves at approximately the same time as the rest of the country.

In Kawasaki City, a state of emergency was declared from April 7 through May 25, 2020, for the first wave of infection, and then from January 8 through March 21, 2021, when another wave occurred. During these periods, a number of requests were made by the national government, including stay-at-home orders, school closings, the closure of large-scale commercial facilities, and reducing the hours of business for restaurants and bars [4]. A state of emergency was also declared from August 2 through September 30, 2021 to respond to the delta wave of infection. Priority measures to prevent the spread of disease were also implemented from April 12 through August 1, 2021 and from January 21 through March 21, 2022 to respond to the alpha, delta, and Omicron waves of infection, respectively. Though these infection control measures and the nationwide stay-at-home orders certainly had an impact on infection rates, COVID-19 vaccinations, which began in February 2021, had a larger role in controlling the later epidemic waves— the effects of which were acutely felt after the delta wave in August 2021.

Normally, the Kawasaki City Infectious Diseases Surveillance Center collects epidemiological and medical information on infectious diseases in Japan and abroad and cooperates with the PHCs to provide training and support for active epidemiological studies and surveys. During the COVID-19 pandemic, from January 1, 2020 to February 5, 2022, the Kawasaki City Infectious Diseases Surveillance Center was consulted for a total of 513 survey requests and case inquiries, including field investigations. Of these 513 inquiries, a total of 220 (43%) were related to surveying COVID-19 clusters, 130 (25%) were related to medical and epidemiological findings totaled, and 110 (21%) were related to individual case consultations, accounting for roughly 90% of the total. Of the total 220 cases corresponding to clusters, 52 (24%) were in medical institutions, 45 (20%) were in elderly care facilities, and 36 (16%) were in facilities for the disabled. Additionally, of the 220 cluster inquiries, 42 (19%) required on-site surveys or meetings with multiple stakeholders, 35 of which were at medical institutions.

When conducting on-site surveys, workers wore surgical face masks even if there was no face-to-face contact with an infected person in the facility, minimized contact with the facility's equipment, and disinfected equipment with alcohol if contact was made. One of the major problems identified in elderly care facilities was poor knowledge of proper measures to reduce the risk of COVID-19. Many resources were devoted to unnecessary tasks that had a minimal impact, such as wearing unnecessary personal protective equipment, installing vinyl curtains, and disinfecting the ceilings and walls of facilities. In response to this, efforts were made to inform facility staff of the simplest possible approach for reducing the risk of infection, review and reinforce specific zoning and cohorting practices, and tried to eliminate concerns about the response measures (Table 1.1). In some facilities, the spread of the infection was thought to have occurred because of poorly ventilated staff changing rooms and restrooms. While it was difficult to improve these infrastructural considerations immediately, solutions were proposed for remedying them, especially when considering the consequences that would come with needing to close the facility in the event of an outbreak.

Since medical or epidemiological evidence is required for the public health response to infectious disease outbreaks, it is important to collect this information and use it to promote evidence-based decision-making. In Kawasaki City, the number of reported cases, as well as the number of requested case investigations, changed according to the epidemiologic period of the pandemic; and as the number of inquiries increased, so too did the reported cases. Put another way, the case counts and case investigation requests increased during each wave of the pandemic but decreased once localized outbreaks subsided. However, once the facilities grew accustomed to the amount of work required for case investigations, this burden of work remained relatively constant and did not always align with the number of cases reported. For instance, the number of cluster responses in Kawasaki City increased during the epidemic waves, but the number of medical and epidemiological inquiries remained high from March to July 2020 and from May to August 2021, which was consistent with insufficient testing systems and the timing of the emergence of variants.

TABLE 1.1 Checklist and associated tasks for the response to nosocomial clusters of COVID-19.

1. Understand the size of medical institutions
2. Identification of the source of infection
   a. Determine whether infection is nosocomial in nature
   b. Check the behavioral history of patients and contacts, and any events occurred
   c. Investigate the possibility of spread from the index patient
3. Implementation of infection control measures
   a. Confirm infection control and isolation status in the hospital (i.e., entry and exit to rooms and wards with minimal need)
   b. Reinstatement of basics of infection control measures
   c. Thorough zoning
   d. Improve support systems tailored to the situation, suspension of hospitalization and discharge, and ward closure
   e. List patients discharged from the hospital (i.e., to prevent subsequent community spread)
4. Confirmation of the existence of any severe or critical patients
   a. Understand the presence of high-risk patients, such as immunocompromised host
   b. Remove susceptible patients or staff (i.e., examining vaccination history)
5. Create a case list
   a. Review history of diagnosis, definitive test methods, and confirmation of progress (e.g., line listing/Gantt chart, epidemic curves, etc.)
6. Hypothesis (How did the cluster spread?)
7. Implement countermeasures meetings, information sharing, and information disclosure
   a. Conduct first conference/meeting as early as possible
   b. Procure necessary items (e.g., test reagents, PPE, etc.)
   c. Review of relevant publications
8. Implement appropriate response to end hospital-acquired infections of COVID-19
9. Review response and challenges
   a. Review response procedures to avoid future clusters of disease

The number of cluster-responding cases in eldercare facilities increased in May 2020 and the period from December 2020 to February 2021, coinciding with a period when the number of clusters in eldercare facilities increased due to larger COVID-19 epidemics. Since April 2021, the support to eldercare facilities has rapidly decreased, suggesting that it is related to the start of vaccination. In Kawasaki City, vaccination rates were bolstered using large vaccination campaigns focused on the staff members of the elderly care facilities, a priority vaccine group, through the development of a campaign in which vaccination teams came to the elderly care facilities and inoculated facility residents with COVID-19 vaccines. When outbreaks occurred in facilities where staff and residents had not been vaccinated, the facility manager, facility physician, health center infectious disease control personnel, vaccination personnel, and LPHI physicians conducted web conferences to plan vaccination campaigns. The necessary number of vaccines were then immediately transported to the facility and vaccinations were administered to facility residents—helping to quickly bring the outbreak under control.

In August 2021, a cluster occurred at an employment and independence support center where infection control was difficult, requiring a lot of time to respond. These centers are residential facilities established to support the employment and independence of persons experiencing homelessness. These facilities are not usually deeply involved in infectious disease epidemics, so outbreaks of COVID-19 posed a unique challenge, as it was difficult to promote cooperation and compliance with residents and staff members generally did not have existing knowledge of response measures. Since there were no existing response procedures in place, responses to outbreaks in these facilities were relevant departments held meetings early in the outbreak to discuss how to appropriately respond. In addition, physicians from the Kawasaki City Infectious Disease Surveillance Center held trainings on infection control, epidemiological investigation, and response methods and procedures, for the staff. These responsibilities were soon left to the staff—quickly minimizing the spread of disease in the facilities and bringing the outbreak under control.

Throughout the COVID-19 pandemic, there have been many inquiries relating to the medical and epidemiological findings from the PHCs, medical institutions, and facilities. A majority of these inquiries were related to the medical system, recommendations for dealing with close contacts and monitoring their health, responses to clusters of disease, risk factors for severe illness and considerations for immunocompromised individuals, the risk of reinfection, COVID-19 variants, drugs and therapeutic agents, personal protective equipment, the availability of facemasks, manuals and guidelines, addressing vaccination errors and needle stick accidents, and conducting infant health examinations.

Developing laboratory diagnosis capacities

COVID-19 was identified in China on January 9, 2020 and soon realized to be caused by a novel coronavirus, SARS-CoV-2. On January 11, the World Health Organization published the whole genome sequence of the virus, which allowed for the NIID to quickly design and prepare primers for polymerase chain reaction (PCR) testing and distribute them to LPHIs nationwide. Because of this, PCR testing for COVID-19 was rapidly available throughout Japan. On January 14, 2020, the NIID and LPHIs carried out the first PCR tests for COVID-19, and the first confirmed case in Japan was announced two days later, on January 16. Over the coming weeks, the NIID continued to refine and improve testing methods and sent necessary laboratory reagents and materials to the LPHIs. By using the national budget to distribute the laboratory testing reagents to LPHIs nationwide, a uniform testing and surveillance system was quickly established for use by local governments, nationwide. Along with the NIID, the LPHIs have capitalized on their specialized knowledge in pathogen testing, and they have worked together to prepare new guidelines and manuals for testing methods.

Since the first case of COVID-19 was reported in Japan, the Kawasaki City Institute for Public Health independently and proactively designed and obtained primers in advance to develop a system that enabled them to conduct their own testing. The primers, probes, and positive controls ordered by the NIID arrived at the Kawasaki City Institute for Public Health on January 30, and a specimen of the index case in Kawasaki City was brought to the laboratory on the next day. However, fortuitously, because the NIID reagents had already arrived, the laboratory was able to conduct testing in a fashion consistent with that being used across Japan.

Once this initial laboratory capacity had been established, the number of examinations increased rapidly with the first epidemic wave and soon threatened to exceed the capacity provided by the two existing real-time PCR machines. As a result, several efforts were taken to augment the existing capacity. The Kawasaki City Institute for Public Health purchased an additional PCR machine and borrowed another to bolster its internal capacity. Even before the COVID-19 pandemic, the Kawasaki City Institute for Public Health had been routinely conducting PCR testing for infectious diseases. But with the existing infrastructure, capacity was limited to testing around 50 specimens per day due to personnel and equipment constraints. However, with these previously mentioned capacity improvements, it is now possible for the institute to conduct more than 300 tests per day.

The city also sought to develop diagnostic capacity by capitalizing on external resources and began administrative inspections at some private facilities on April 1 to strengthen the laboratory system. Relatedly, new testing sites commissioned by the government were opened on May 11. As a result, by June 2020, laboratory specimens could be tested even in clinics, and the number of tests increased significantly allowing for the detection of many more cases of COVID-19.

While these increases in testing capacities were an important first step, they also exposed other challenges. For example, as the number of laboratories performing tests increased, it soon became apparent that there was no system in place for collecting specimens from close contacts in the event of clusters in elderly care or childcare facilities. To address this gap, Kawasaki City relied on local field epidemiologists to bolster specimen collection. Kawasaki City previously dispatched city health officials to participate in an introductory Field Epidemiology Training Program (FETP) sponsored by the national government. This training was followed by an additional, supplemental training, FETP-Kawasaki (FETP-K) —providing officials with at least two years of on-the-job training to support the response to infectious disease outbreaks [7]. In response to the COVID-19 pandemic, a sample collection team, comprised mainly FETP-K members, was dispatched to the facilities in June and July to collect specimens in the facilities and carry out examinations before a formal sample collection and testing system was established for these facilities.

As the pandemic has grown and evolved over time, the Kawasaki City Institute for Public Health has shifted its focus to providing testing for genome sequencing of variants or clusters that require rapid responses, while more routine testing has been commissioned to private laboratories. This change has been accompanied by a shift in the role of the institute, from one focused on testing to one more focused on genomic analysis and variant surveillance. The institute had introduced next-generation sequencers prior to the COVID-19 pandemic, but the pandemic has resulted in many staff members acquiring the skills needed to perform genomic analyses on a daily basis. It has been difficult, however, to reliably secure the reagents and consumables required for genomic sequencing, especially those reagents and consumables that are manufactured outside of Japan.

More recently, laboratories in Japan have introduced fully automatic genetic test equipment, antigen quantitative test equipment, and antigen qualitative kits, making it possible to conduct testing at general medical facilities, clinics, and even in nonmedical facilities. The simplification of diagnosis has drastically reduced the burden of testing itself at the Kawasaki City Institute for Public Health, but the rapid increase in the number of cases has made it difficult to collect accurate epidemiological information, as epidemiological surveys at PHCs are not always available. This has proven to be a formidable challenge, as it is necessary to analyze current and accurate data to the greatest extent possible to make informed recommendations for reducing the risk of COVID-19.

Vaccines and mass vaccination exercises

On October 23, 2020, the National Ministry of Health, Labour and Welfare issued a notification to each municipality that the "Preparatory Guidelines for Ensuring the COVID-19 Vaccination System" would be applied and preparedness projects implemented [8]. The purpose of these guidelines was to gradually develop the system necessary for the mass distribution of vaccines, even before the vaccines themselves were available. As specified by the guidelines, municipalities were responsible for contracting with medical institutions, providing vaccine recommendations to inhabitants, paying inoculation costs, and providing individual notifications (examination slips, coupons), while prefectures were responsible for coordinating with local wholesalers and coordinating with municipalities. Still, this proved challenging as the specific characteristics and effects of the vaccines were not known at the time, and uncertainties regarding the development, production, and purchasing of vaccines made it difficult to determine when vaccinations would begin.

In addition to these issues related more to vaccination systems, the first available vaccines for COVID-19 were intramuscular injections using a novel mRNA technology. Thus in addition to collecting the scientific evidence necessary for proving the efficacy and safety of vaccines, it was also necessary to ensure that the media was disseminating factual information regarding the vaccines, as there was lots of information circulating—some that were accurate, and some that were not.

As time went on, and in anticipation of the arrival of vaccines, the Immunization Act and the Quarantine Act were proactively revised on December 2, 2020 to bolster the implementation of mass vaccination campaigns [9]. Because COVID-19 vaccinations would rely on a different system than that used for routine vaccinations, an exception for temporary vaccinations was established. Further, provisions of the Immunization Act were applied for the reporting of suspected adverse reactions to vaccines, and the government, not pharmaceutical companies, was responsible for the health damages that may have been caused by vaccines. The Pfizer-BioNTech COVID-19 vaccine was exceptionally approved in Japan on February 14, 2021. Shortly thereafter, on February 16, the Ministry of Health, Labour and Welfare issued a notification directing medical facilities to being vaccination efforts against COVID-19, and vaccination campaigns began in earnest on February 17, 2021.

Kawasaki City has a history of conducting mass vaccination exercises, primarily in preparation for pandemic influenza. Similarly, on December 27, 2020, prior to the introduction of COVID-19 vaccines, Kawasaki City conducted a training exercise for COVID-19 vaccinations, sponsored by the Ministry of Health, Labour and Welfare, and in cooperation with Pfizer and BioNTech [10].

Because so many individuals needed to receive the COVID-19 vaccination, it was necessary to not only consider individual vaccinations but also consider and develop systems that enabled mass inoculation. Still, in Japan, vaccines are normally conducted on an individual basis in cooperation with medical institutions, and mass vaccination systems are not the norm. Mass vaccinations for influenza stopped in 1994 and apart from conducting mass vaccinations for polio, tuberculosis, and measles in times of emergency, there have been limited opportunities to conduct large-scale mass vaccination campaigns. Therefore it was deemed necessary to implement training—provided by the Ministry of Health, Labour and Welfare to municipalities across Japan—on the establishment and operation of mass vaccination venues by PHCs to safely, reliably, and rapidly vaccinate the population against COVID-19.

As a part of this training, volunteers were recruited from the general public to act as patients, while health care personnel gained experience and practiced the actions required for implementing a mass vaccination campaign in the settings in which they would actually be conducted. This training also allowed officials to appropriately consider other important elements, such as the time required, personnel security, vaccine storage, infection control, and safety assurance system, among other things. In Kawasaki City, this training was implemented in two different venues—one at a medical institution operated by the Kawasaki Medical Association and the other at a nonmedical institution located in a university gymnasium—as a means of constructing a practical and versatile vaccine management system. The following paragraphs describe the training that occurred at nonmedical institutions (i.e., the university gymnasium), although a similar training similar was implemented at the venues located at medical institutions.

Because the training and vaccination efforts would occur during the COVID-19 pandemic, preventive measures that minimized the risk of disease transmission were emphasized, including the "Three Cs"—closed spaces, crowded places, and close-contact settings. As a result of these considerations, it was decided that the venue would contain the following booths: (1) a reception booth outside the venue, (2) a reception booth inside the venue, (3) a booth to fill in the medical examination form, (4) a booth to confirm and review the medical examination form, (5) a waiting area for the medical examination, (6) medical examination rooms, (7) a waiting area for vaccinations, (8) vaccination rooms, and (9) rooms for issuing certificates of inoculation, (10) postvaccination follow-up rooms; (11) emergency first aid rooms, and (12) vaccine dilution rooms (Fig. 1.1). This format allowed for the venue to avoid closed spaces, crowded places, and close-contact settings

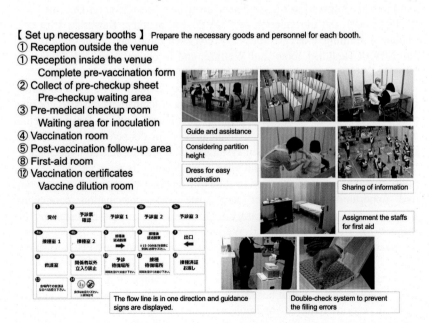

FIGURE 1.1 Summary of COVID-19 vaccination training provided by the Kawasaki City COVID-19 Vaccination Site Operational Training Report.

to the greatest extent possible. Responsibilities for those staffing each element were determined, and a manual with frequently asked questions was prepared to aid workers.

As an infection countermeasure, those working at the vaccination venues were encouraged to check their health in advance, wear a facemask, and practice good hygiene, including frequently disinfecting their hands. In addition to wearing masks, workers were also instructed to wear gloves and gowns when giving vaccines or when physicians and nurses were performing medical procedures. While there were various opinions on the benefits and drawbacks of the personal protective equipment worn by the staff, efforts were taken to ensure that the equipment was as unobtrusive as possible and made it easy to perform the necessary tasks while minimizing the risk of infection.

Those who were receiving vaccines had their temperatures measured at the entrance upon arrival, were provided facemasks to wear throughout their time at the venue, and were offered access to hand sanitizer through sanitizing stations that were distributed throughout the venue. Additionally, vaccine recipients moved in a unidirectional line and were instructed to maintain appropriate spacing as a means of keeping the environment well-ventilated and reducing the risk of infection in the venue. When setting up the venues, staff was assigned to each booth, booths were labeled, and the pathway that patients should follow was displayed within the venue.

The training initially sought to determine how many individuals could be vaccinated in 30 min. To measure this, the total number of people who arrived at the postvaccination follow-up rooms was recorded, as were the times when patients arrived at each booth. The total time spent at the mass vaccination venue was then determined to be the amount of time required for an individual to move from the check-in booth to the follow-up observation booth, plus 15−30 min for follow-up observations. Results from the training showed that 15 individuals were able to complete the inoculation within the 30 allotted minutes, and the average time required per person was approximately 15 min. The results also suggested that potential bottlenecks included completing the preexamination questionnaire and completing preliminary examinations, so these elements were reviewed to promote efficiency. Based on training results, it was decided to confirm the roles of physicians and staff in the emergency first aid rooms and to create a manual for these.

At the same time as this training was conducted, additional training on handling the vaccine was also provided to staff. This training emphasized how to receive the vaccine, how to properly store it in freezers, how to dilute the vaccine, and how to fill a syringe. As a means of ensuring patient safety, it was determined that the dilution of the vaccine and syringe filling should be conducted using a double-check system whereby two different people reviewed these steps.

After reviewing the time and personnel required to stand up the mass vaccination venue, it was determined that, when compared to individual vaccinations, the mass vaccination venues required more resources and staff but were also able to inoculate greater numbers of people. For this reason, the municipal government decided to develop a system that allowed for mass vaccination venues and individual vaccination to be implemented at the same time.

After resolving the detailed operational problems that were identified during the training, several mass vaccination venues were established in Kawasaki City in April 2021 and began distributing vaccines shortly thereafter. Over time, as staff became more familiar with the required tasks, the time required for vaccinations was reduced, allowing for mass vaccinations to be conducted in an efficient fashion. However, this progress was also accompanied by other unanticipated problems such as vaccine loss, storage error, and erroneous inoculation; these in addition to addressing the broader problem of how to raise the vaccine coverage rates in the face of widely circulating mis- and disinformation regarding their safety and efficacy. While the training was very effective for preliminary preparations and establishing the vaccination system, it was soon realized that greater attention needed to be given to contingencies, including reactions after inoculation, such as anaphylaxis. Furthermore, it is essential to continuously collect necessary information, such as the scientific rationale for the efficacy and safety of vaccines and to provide the correct information to the public.

Responding to the first case of the Omicron variant in Kawasaki City

The Omicron variant resulted in a major epidemic wave of infection in Japan. Throughout the pandemic, airport screening and quarantine capacities have been strengthened as a response measure. Beginning on May 31, 2022, a policy was implemented in which specimens from travelers are collected upon entry into Japan; should an individual test positive for COVID-19, the NIID works to identify variants from laboratory-confirmed COVID-19 patients and these patients with positive Omicron variants are isolated at medical institutions. The first imported case of the Omicron variant was detected in Japan on November 30, 2021. By December 15, more than 30 additional cases had been detected through airport screenings, but none of these cases were considered to be locally acquired.

Still, on December 16, 2021, health authorities in Tokyo announced that they had identified a person who was infected with the Omicron strain in the metropolitan area. The individual was a woman who lived in the area and had recently returned from the United States but tested negative upon arrival at the airport. While staying at home, she developed symptoms, tested positive using PCR testing, and subsequent genomic analysis confirmed infection with the Omicron variant. The previous day, on December 15, a man who had been in close contact with her also tested positive for COVID-19, but this individual did not have any international travel history. Both of these individuals had received two vaccinations for COVID-19.

It was soon discovered that the man had gone to work before his infection was realized and that he had attended a large soccer match with his family at the Todoroki Athletics Stadium in Kawasaki City on December 12. As such, there was concern that he had been infected with the Omicron variant when he attended the soccer match and that the strain would spread throughout the city. This particular case presented unique challenges, as the man's residence, his place of employment, and the soccer stadium are all located in different municipalities. Because of this, the Ministry of Health, Labour and Welfare, the NIID, authorities from Tokyo, authorities from Kawasaki City, and other relevant officials held a series of web conferences on December 15 to share information and determine what actions needed to be taken. Subsequent genomic analysis confirmed that the man was, indeed, infected with the Omicron variant.

The Todoroki Athletics Stadium, where the infected man attended the soccer match, is an athletics stadium owned by Kawasaki City with a maximum capacity of 25,000. After interviewing the man, authorities learned that he and his family were watching the game in sections 216 and 217 of the stadium. Given that there were no other known cases of community-acquired Omicron infections in Japan at the time, there was some initial discussion that everyone who attended the match should be tested as a means of identifying infected individuals. While eating and drinking were permitted in seats at the stadium, it is an outdoor venue, and authorities worked with the Japan Football Association to confirm that the man wore a facemask while watching the game and that loud cheering had been limited. Because of this, rather than testing everyone who attended the match, authorities decided to test only the 75 individuals who watched the game around him, particularly in sections 216 and 217.

To contact those who purchased tickets to attend the soccer match, Kawasaki City officials gathered a list of addresses and names from the Japanese Football Association—the organization that sponsored the event. Of the 75 persons of interest, 39 lived in Kawasaki City and were contacted by telephone by the authorities at the Kawasaki City Institute for Public Health; the other 36 individuals who resided outside of Kawasaki City were contacted by their jurisdictional PHCs to request that they take a diagnostic PCR test. Municipal authorities in Kawasaki City also conducted a communication campaign using the city's website and Twitter accounts to encourage those who attended the match to test for infection, avoid going out as much as possible, and seek medical attention if they did not feel well.

By December 26, 35 of the 39 individuals who lived in Kawasaki City were confirmed to have a negative test, but four of the individuals could not be reached. Similarly, for the 36 individuals living outside of Kawasaki City, 35 were confirmed to have tested negative, but one individual could not be contacted. Still, the individuals who could not be contacted were sitting relatively far away from the infected man, and, by this time, 14 days had passed since the date of the soccer match without additional reports of infected persons. It was, thus, concluded that there was no further risk of infection from this particular event.

Conclusion: lessons learned from Kawasaki City

When compared to other countries, one of the defining characteristics of the COVID-19 response in Japan, thus far, is that the government has not imposed strict limitations on the behaviors of individuals. The pandemic response specific to Kawasaki City has been in line with the national response and has not differed significantly from the responses in other ordinance-designated cities in Japan.

Ultimately, the experiences from Kawasaki City suggest that it was possible to control the COVID-19 pandemic largely through voluntary measures and the self-restraint demonstrated by citizens. However, the pandemic response did not come without its challenges. Responding to outbreaks at facilities presented a challenge, as they often lead to rapid increases in a number of cases and severely ill persons. This was especially true in elderly care facilities, because this demographic group is more vulnerable to infection and more likely to develop severe illness. As demonstrated throughout the COVID-19 pandemic, outbreaks of COVID-19 can cause hospitals to reach capacity, which can prevent patients with more mild diseases from being hospitalized. In these instances, it is of the utmost importance to take the measures necessary for providing care to infected individuals, while also taking infection control measures, to prevent the spread of disease. What is of equal importance is that these standards and guidelines are championed before an emergency so that they may be correctly implemented in the early stages of an epidemic. It is also important to introduce simple response methods that may be conducted even by laypersons.

To implement an effective response, it is also important to collect and share accurate epidemiological and medical information. In reflecting upon the experiences of Kawasaki City, given that many of the inquiries received by the PHC related to medical considerations, establishing an administrative institution that can collect and disseminate specialized knowledge and information could prove beneficial for improving pandemic preparedness and response in the aftermath of the COVID-19 pandemic. This is especially true when considering the growing risk of emerging and re-emerging infectious disease epidemics. Additionally, while manuals and guidelines will be published at the national level as needed and necessary, it is also necessary for each municipality to develop detailed guidance and response standards during normal times so that they are prepared for the response to actual events.

Ideally, the accurate diagnosis of infectious disease will inform and influence the subsequent response, so establishing a standardized, national surveillance system is essential. To this end, as a method of preparing for a health emergency, it is important to develop a system for the rapid collection and testing of specimens, such as sample collection teams. Thus far, these streams of work have been conducted on an ad hoc basis for the COVID-19 response, but these systems should be strengthened in the future as an established, more permanent system of national and local governments—as is the case with the Laboratory Response Network in the United States of America.

It is also critically important to effectively utilize vaccines, which are one of our most effective tools for protecting people from the risk posed by infectious diseases. The vaccines for COVID-19 were newly developed and used relatively early in the pandemic response. The training that was conducted before the vaccines were available was very effective as a preparation measure, but experiences have shown that it is also necessary to consider how responses will deal with unforeseen situations, such as the incorrect storage of vaccines that require ultra-low temperatures or incorrect vaccinations. Furthermore, it is essential to continuously collect important information regarding the vaccines, such as vaccine efficacy and safety, and to promptly provide the correct information in an accessible way to the public.

Finally, in Japan, the development of the epidemiological information system, including medical information, was particularly slow. Because of the rapid increases in COVID-19 cases in various pandemic waves, it has been difficult to collect and input epidemiologic data, primarily because of challenges in data entry. While there are certainly advantages to providing prefectures a degree of flexibility in tailoring their response to their local situations, data collection did not occur in a standardized way across the country. This has presented some unique challenges during the response to COVID-19 because multiple occasionally redundant information collection systems developed, leading to enormous amounts of clerical work for each municipality. Digitizing these information systems could simplify these processes in the future, as they currently require manual work. Doing so would provide for a more nimble and informed response, as during an epidemic or pandemic, it is important to adapt the response based on cleaned, standardized, and analyzed data to the greatest extent possible.

References

[1] Kawasaki City. Foreign resident population by age by jurisdiction, <https://www.city. kawasaki.jp/shisei/category/51-4-3-7-0-0-0-0-0-0.html>; 2021 [accessed 27.09.22].

[2] Government of Japan. Act on the prevention of infectious diseases and medical care for patients with infectious diseases (the Infectious Diseases Control Law). Tokyo: Government of Japan; 1998.

[3] Government of Japan. Act on special measures for pandemic influenza and new infectious diseases preparedness and response. Tokyo: Government of Japan; 2013.

[4] Zheng J. SARS-CoV-2: an emerging coronavirus that causes a global threat. International Journal of Biological Sciences 2020;16(10):1678−85. Available from: https://doi.org/10.7150/ijbs.45053.

[5] National Institute of Infectious Diseases. COVID-19 as of May 2020. Infectious agents surveillance report 2021;41(7):103-5.

[6] National Institute of Infectious Diseases. Situation report of genomic surveillance of COVID-19, <https://www.niid.go.jp/niid/images/cepr/covid-19/220810_genome_surveillance.pdf>; 2022 [accessed 14.08.22].

[7] Misaki T, Saito T, Okabe N. Building a robust interface between public health authorities and medical institutions in a densely populated city: State-of-the-art integrated pandemic and emerging disease preparedness in the Greater Tokyo Area in Japan. In: Katz R, Boyce M, editors. Inoculating cities: Case studies of urban pandemic preparedness. San Diego: Academic Press; 2021. p. 99−114.

[8] Ministry of Health, Labour and Welfare of Japan. Preparatory guidelines for ensuring the COVID-19 vaccination system, <https://www.mhlw.go.jp/content/000714253.pdf> [accessed 27.09.22].

[9] Government of Japan. Act to partially revise the immunization act and the quarantine act. Tokyo: Government of Japan; 2021.

[10] Kawasaki City. Operational training report of COVID-19 vaccination site, <https://www.city.kawasaki.jp/350/cmsfiles/contents/0000124/124018/houkokusho.pdf>; 2021 [accessed 14.08.22].

Chapter 2

COVID-19 pandemic: how Lagos State Government managed the COVID-19 pandemic and the tripod of biosecurity

Akin Abayomi[1,2] **and Tunde Ajayi**[2]

[1]*Division of Haematology, University of Stellenbosch, Western Cape Province, South Africa,*
[2]*Nigerian Institute of Medical Research, Nigeria Ministry of Health, Lagos State Secretariat, Lagos, Nigeria*

Background

Background on Lagos, Nigeria

Lagos is a state in southwestern Nigeria that holds the smallest land area of all Nigerian states, with an area of 3,577.2 square kilometers.[1] The state is further subdivided into five administrative divisions, 20 local government areas (LGAs), and 37 Local Council Development Areas (Fig. 2.1), which, in accordance with the federal structure of the Nigerian government, are tasked with governing the area [1].

Despite Lagos' relatively small land area, it is the second most populous state in Nigeria. According to the last state population census, conducted in 2006, the estimated total population of the state was some 9.1 million inhabitants [2]. And, with a growth rate of 3.2%, the population is currently estimated to be 15.1 million people and reach 19.4 million people by 2030.[2] This high growth rate is fueled by frequent rural-to-urban migration, with people pursuing the economic and educational opportunities that exist within the

---

1. This includes an estimated 2797.72 km^2 of land area and 779.56 km^2 of water area.
2. Estimates from the United Nations suggest that the population may be higher—with an estimated 13.5 million inhabitants in 2018, and a projected 20.6 million inhabitants by the year 2030 [3]. Regardless, it is widely accepted that the population of the larger metropolitan region is larger than 10 million persons, making Lagos a mega-city.

Inoculating Cities, Volume II: Case Studies of the Urban Response to the COVID-19 Pandemic.
DOI: https://doi.org/10.1016/B978-0-443-18701-8.00002-3

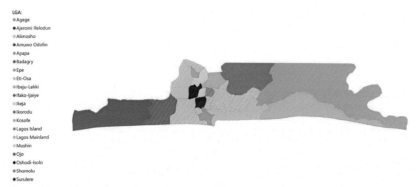

LGA:
- Agege
- Ajeromi Ifelodun
- Alimosho
- Amuwo Odofin
- Apapa
- Badagry
- Epe
- Eti-Osa
- Ibeju-Lekki
- Ifako-Ijaiye
- Ikeja
- Ikorodu
- Kosofe
- Lagos Island
- Lagos Mainland
- Mushin
- Ojo
- Oshodi-Isolo
- Shomolu
- Surulere

FIGURE 2.1   The local government areas of Lagos State.

state.[3] As such, the state is densely populated, and the human population density in Lagos State stands at approximately 5,419 people per square kilometer—far above the national average of 195 people per square kilometer. Much of this population resides in the metropolitan city of Lagos, which accounts for 37% of the land area of the State, but 85% of the population [1].

Lagos has acted as the principal entertainment, industrial, and political center of Nigeria over time [1]. Lagos formerly served as the capital of Nigeria, and while it no longer fills this role after the capital was moved to Abuja in 1991, Lagos is an economic hub and financially viable state— accounting for roughly 27% of Nigeria's total gross domestic product, 65% of commercial activities in the country, and generating the highest internal revenue of all Nigerian states [1]. Further, the commercial attributes of Lagos render it the most likely epicenter of any imported pathogen of high consequence because the city accounts for over 60% of the air traffic into Nigeria and has two seaports, one international land border, and several internal road inlets from neighboring Nigerian states. The city also imports a large number of livestock on a daily basis to feed its large population.

In the last 10 years, the Lagos State Government has made tremendous progress in terms of rapid economic growth, improved infrastructure and services, and a significant reduction in crime rates. This has provided an enabling environment for millions of Lagos state inhabitants to find their way out of poverty. The state government has also made great strides in its quest to increase value for money in public spending, improve the business climate in Lagos, maintain fiscal sustainability, and properly monitor and manage financial and health risks.

---

3. Many of these individuals are unskilled laborers, which has resulted in the rapid expansion of slum communities within the city.

Background on the public health system in Lagos

Nigeria has a pluralistic health care system with both modern and traditional health care professionals, as well as governmental and private health care facilities. The three layers of government are jointly responsible for providing health care. Nigeria's three levels of government—the Federal Ministry of Health, the State Ministry of Health, and LGAs—operate a decentralized health care system. The Federal Ministry of Health is responsible for the coordination and implementation of national health policy and the LGAs are expected to have oversight of the primary health services through primary health care centers. In 2018 there were an estimated 297 primary health centers and 1,187 beds in Lagos State; there were also some 258 doctors, 672 nurses, and 1,824 other health care workers (Table 2.1) [4]. There are also currently 25 accredited private laboratories and 4 public health laboratories in Lagos State.

Lagos State, through its Ministry of Health, primarily provides oversight over all three levels of health care (i.e., primary, secondary, and tertiary health care). Additionally, through the State Primary Health Care Development Agency, the Ministry of Health manages the implementation of primary health care at the LGA level. Aside from leading the execution of the AIDS, tuberculosis, and malaria program interventions, the Ministry of Health also adapts national policies and strategies and oversees other high-priority health-related initiatives.

The Lagos State Government is supported by the private sector to provide much needed health services. According to the Lagos State Health Regulatory Agency, there are more than 3,000 private health care facilities

TABLE 2.1 Estimated health care system capacities in Lagos, 2018.

Health care system capacity	Total number	Number per 100,000 persons [a]
Primary health centers	297	2.2
Beds	1,187	8.9
Personnel	2,754	20.7
Doctors	258	1.9
Nurses	672	5.1
Other	1,824	13.7

[a]These estimates were produced using a population of 13.3 million, which is the estimated population of Lagos in 2018 using the estimates from the 2006 census and assuming an annual growth rate of 3.2%.
Source: Data are sourced from the 2019 Lagos Bureau of Statistics' Abstract of Local Government Statistics [4].

operating in Lagos comprising large multistorey medical complexes to simple back street clinics. Lagos also has a health insurance scheme managed by the Lagos State Health Management Agency to provide its residents with access to quality and equitable health care services while ensuring financial protection. Further, there is also a plethora of traditional medicine practitioners providing a variety of services from traditional birth attendants, bone setters, herbalists, and spiritualists; more exotic practices such as acupuncture, aromatherapy, and homeopathy are also growing in popularity.

Lagos State's approach to emergency preparedness

Perhaps no disease has had as great an impact on Lagos State's approach to emergency preparedness as Ebola virus disease (EVD). On July 20, 2014, an acutely ill individual traveling from Liberia arrived at Lagos Murtala Muhammed International Airport. This individual was immediately transferred to a private hospital in Lagos, and Lagos University Teaching Hospital reference laboratory was used to confirm EVD infection on July 23, 2014. A rapid response was implemented in Lagos and by August 20, 2014, all contacts of cases and potential cases had been identified and contacted. Fortunately, this rapid response prevented the outbreak from escalating and two months later, on October 20, 2014, the World Health Organization (WHO) declared Nigeria to be Ebola-free.

Governance and policy

Following this EVD outbreak, Lagos State recognized the need to create a preparedness strategy that could be implemented in the event of any subsequent outbreaks. In November 2014, a committee known as the State Ebola Viral Disease Research Initiative (SEVDRI) was formed. SEVDRI was tasked with developing a framework for emergency preparedness that would improve the State's resilience and ability to respond to public health risks, including biological ones.

As a part of this mission SEVDRI, working as a technical partner with the Global Emerging Pathogen Treatment (GET) Consortium,[4] concluded that in addition to effective public health and surveillance systems, there was a need to incorporate emergency preparedness measures into legislation and policies. Therefore an evaluation of Lagos State's existing laws, rules, and regulations was conducted to inform and direct the creation of stand-alone emergency preparedness, biosecurity, and biobanking law and to suggest and seek revisions to current legislative acts pertinent to biosecurity, if applicable. This evaluation gave special consideration to the domestication of

---

4. The GET Consortium is an African-led, international, multidisciplinary collaboration that was established in 2014 in response to the Ebola virus disease outbreak in West Africa.

elements of leading international frameworks and agendas such as the International Health Regulations (2005), Practices of Veterinary Services, the Biological Weapons Convention, the United Nations Security Council Resolution 1540, and the Global Health Security Agenda. Results from this effort revealed that, at the time, there were no laws or regulations to support biobanking, emergency preparation, or the dangers that contemporary biotechnology poses to the security of the world's health, such as the creation, manufacture, stockpiling, or deployment of biological weapons.

SEVDRI then changed its name to the Biosafety, Biobanking, Biosecurity, and Biocontainment (B4) Committee in June 2015 and adopted a comprehensive and multifaceted strategy to: (1) identify legislative loopholes in the State's disaster preparation, biobanking, and biosecurity laws in collaboration with the Lagos State legislature; (2) create the Lagos State Biosecurity and Biothreat Reduction Road Map, a framework and plan for implementing biobanking and biosecurity in Lagos State concerning readiness and quick reaction; and (3) create the Lagos State Biosecurity and Biobanking Governance Council (LSBBGC), an association charged with helping the State protect its biological space to ensure its health and that knowledge derived from it is sovereign and protected. Importantly, the LSBBGC is required to consider a diverse range of expertise and experts in agriculture, anthropology, biobanking, biosecurity, civil society, environmental sciences, ethics, law, molecular biology, public health, and security, which are among the group's 12 members. These actions, especially establishing the LSBBGC, made it possible to scale up emergency preparations against novel infectious diseases.

The B4 Committee also developed a policy on emergency preparedness, biobanking, and biosecurity, which underwent multiple rounds of review with stakeholders before being approved by the LSBBGC and the critical stakeholder ministries. The policy drew on the State's existing public health infrastructure, including the Lagos State Emergency Management Agency, Lagos State Ambulance Service, Lagos State Waste Management Authority, Lagos Mainland Infectious Disease Hospital, the State Environmental Health Monitoring Unit, and the Integrated Disease Surveillance and Response system at both the State and Local Government levels for routine surveillance and notification. The GET Consortium, serving as biosecurity consultants to the Lagos State Government, also assisted in this procedure.

Another notable policy effort was the development of a five-year biosecurity policy and plan, published in 2018. The revised policy drew heavily on knowledge accumulated from combating previous crises and epidemics, such as EVD in 2014.

For instance, in response to the 2014 EVD epidemic, an incident management center was created in Lagos State by the Nigeria Centre for Disease Control and the Lagos State Ministry of Health. This incident management center, which was subsequently redesignated as the National Emergency

Operations Center (EOC) the same year, served as the implementation arm of the country's response to the EVD epidemic. The EOC used an incident management strategy to quickly recognize and address disease outbreak concerns and included six units: (1) epidemiology and surveillance, (2) communication and social mobilization, (3) case management and infection prevention and control (IPC), (4) laboratory services, (5) point of entry (i.e., port health), (6) management and coordination. The new policy, however, increased the number of units to nine by adding research, logistics and supply (including the distribution of personal protective equipment), and case management independent from IPC (Fig. 2.2).

Included in the B4 Committee was the EVD Research Committee, which had been formed during the EVD response. This Committee was assigned the duty of managing the construction of the emergency preparedness infrastructure, including a biosafety level-3 (BSL-3) laboratory and a biobank facility, now known as the Lagos State Biobank. These projects were funded by the Lagos State Government and the Government of Canada and are notable for several reasons. At the most fundamental level, they resulted in Lagos State becoming Nigeria's first state to develop and build a BSL-3 laboratory. This laboratory currently represents the sole operational BSL-3 laboratory in Nigeria and, at the time of writing, maintains the ability to conduct 1,500 polymerase chain reaction (PCR) tests each day. But beyond this, the biobank is notable because it is the first energy mix hybrid biobank in West Africa—principally powered with solar energy sources, which have been deemed an ecologically friendly energy source.

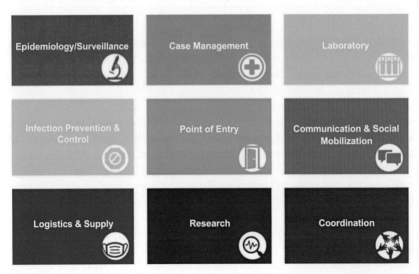

FIGURE 2.2 The nine pillars included in the Lagos State Emergency Operations Center.

A director, a quality assurance manager, information technology employees, a risk assessment officer, an infectious disease specialist (i.e., a virologist), and 12 other persons work for the biobank. These individuals form the backbone of the GET Consortium-supported Biobanking and Biosecurity team. Following the 2014 EVD outbreak, advocacy efforts persisted and resulted in significant funds being invested to rapidly institutionalize emerging infection, prevention, and control procedures. These investments included the creation of standard operating procedures for personal protective equipment (PPE) and collaborating with the Federal Ministry of Health's Port Health Services to establish protocols for screening individuals entering Lagos from its land, sea, and air borders, beginning in April 2017.

The Lagos State Ministry of Health also designated the Lagos Mainland Infectious Disease Hospital as the referral center for suspected or confirmed cases of emerging infectious diseases. The government also mandated the establishment of a particular isolation unit for suspected cases of EVD and Lassa fever, which represented the infectious disease of concern at the time and provided a buffer supply of PPE for medical staff at the hospital.

Coordinating systems

The LSBBGC was entrusted with overseeing and coordinating the various biosecurity measures, including the Emergency Preparation and Biosecurity Programme. Throughout these efforts, the LSBBGC has maintained the responsibility for making decisions regarding key technical and governance concerns. In consultation with the LSBBGC, and considering the established emergency preparedness and biosecurity policies, guidelines, and frameworks, the Lagos State Government established Epidemic Preparedness and Response Committees and Rapid Response Teams (RRTs) in each LGA through the Directorate of Epidemiology, Biosecurity, and Global Health.

The responsibilities for creating and supervising the execution of emergency preparation strategies, action plans, and procedures were then delegated to the Epidemic Preparedness and Response Committee in each LGA, which comprised the manager of the local government council and the LGA health team. More specifically, the Medical Officer of Health, the Apex Chief Nursing Office, the Apex Community Health Officer, the Apex Monitoring and Evaluation Officer and Assistant, the Local Immunization Officer, and the Disease Surveillance and Notification Officer were all members of the LGA health team. Per recommendations from the Federal Ministry of Health, the RRTs are technical and multidisciplinary teams that are readily available for swift mobilization and deployment in times of crisis. The RRTs in Lagos LGAs consist of the chairman of the LGA, the head of the Community Development Committee, the head of the Neighborhood Security Commission, one divisional police officer, the head of works and housing, and every member of the LGA health team.

Capabilities

After the 2014 EVD epidemic was successfully contained, Lagos State Government began the process of strengthening its ability to handle and mitigate upcoming biosecurity risks. Accordingly, in partnership with the GET Consortium, the state government organized five biosecurity conferences in the West African subregion between 2015 and 2020. It also provided key personnel with training in biosecurity and biothreat reduction.

The LSBBGC and the Biobanking and Biosecurity Team also sought to ensure that the Lagos State Biobank complied with global standards. This was accomplished by employing all the governance mechanisms previously created and by partnering with various genomics and bioinformatics consortia including H3Africa, B3Africa, and the GET Consortium. Several assessments were also completed. These included capability assessments, which included work to compare the Lagos State Biobank to the biobank at the University of Stellenbosch in South Africa, as well as conducting a standard risk assessment with support from the WHO to look for microbiological organisms in bodies of water. A risk assessment plan that utilized a One Health strategy was also created.

Numerous employees of the Lagos State Ministry of Health have also participated in training to develop a workforce equipped for the response to public health emergencies. These have included certification training and seminars on emergency preparedness, biosafety, biosecurity, infection prevention and control, data governance, and sample governance. The GET Consortium, WHO, Public Health England, South African National Biodiversity Institute, Stellenbosch University, and Western Cape University have facilitated these training sessions, which have greatly strengthened the human resources for health in the state.

Still, other efforts have focused more on surveillance, forewarning, and information management. Since the conclusion of the EVD epidemic, the previously discussed EOC has served as a focal point for planning and coordinating action during public health emergencies. It has also served as a model for other EOCs that have been established in Nigeria.[5] But in Lagos State, the Epidemiology and Surveillance Unit has been bolstered and strengthened through the creation of more robust surveillance structures. These efforts have included Lagos State Government-funded and WHO-supported sessions that provide a forum for discussing difficulties associated with the surveillance and reporting systems. They also include monthly meetings where LGA Disease Surveillance and Notification Officers discuss surveillance operations in their respective LGAs.

---

5. As of 2019, there are a total of 20 EOCs in Nigeria.

Resources

The biobanking plan and the emergency preparedness initiatives have received sufficient funding from Lagos State. This financing has supported several capacity-building initiatives, infrastructure improvements, frequent meetings, and a robust surveillance and reporting system. Foreign governments and organizations have also contributed financing for preparedness initiatives in Lagos. For instance, as mentioned previously, the Lagos State Biobank was developed through a collaboration between the Lagos State Government and the Government of Canada, which helped fund the US$4.5 million initiative.

COVID-19 in Lagos

Nigeria was classified by the WHO as one of 13 African countries with a higher risk of spreading COVID-19. On January 28, 2020, the Nigerian Centres for Disease Control worked with 22 Nigerian states, including Lagos State, to activate state-level EOCs and link them with the National Incidence Coordination Centre. Three days later, on January 31, 2020, the Federal Ministry of Health activated the national EOC and, in conjunction with the WHO, set up the Multi-sectoral Coronavirus Preparedness Group, which was led by the Nigerian Centres for Disease Control. The Ministry of Health also developed a data collection, analysis, and interpretation system to complement each EOC pillar. These efforts were enhanced by investing in technology to automate the data collection system through the initiation and implementation of the Lagos State Emergency Response System, which allowed for authorities to access reliable, real-time data.

The first COVID-19 case in Nigeria was confirmed by the Federal Ministry of Health on February 27, 2020, making Nigeria the third country in Africa to do so after Algeria and Egypt. The index case was an Italian national who, on February 24, 2020, took a flight from Milan, Italy, to Lagos, Nigeria, and then drove the same day to his company's location in Ogun State, which borders Lagos State to the north. He presented symptoms consistent with COVID-19 on February 26, 2020 at the company clinic and was subsequently sent to the infectious disease hospital in Lagos. The next day, real-time reverse transcription PCR was used to confirm the diagnosis of COVID-19.

Other passengers on the February 24th aircraft were among 216 contacts in Lagos and Ogun States designated for a 14-day follow-up, with 40 of these contacts further designated as high-risk. The country's second case of COVID-19 was identified 11 days later—an asymptomatic contact of the index case. Since that time, the epidemiology of COVID-19 in Nigeria has changed, and cases have been identified in 35 of the country's 36 states, as well as the Federal Capital Territory.

The response to COVID-19 in Lagos

The response to the COVID-19 pandemic in Lagos can be divided into five periods—pre-index case, post-index case, and three pandemic waves—each featuring unique objectives. More specifically, the primary objective of pre-index case period was to anticipate and prepare for cases of COVID-19, the objective of the post-index case period was to mitigate the spread of the SARS-CoV-2 virus, the objective of the first wave was to adapt the health system to manage the pandemic, that of the second wave was to learn and strengthen health systems while reducing deaths, and the objective of the third wave was to pursue resilience and herd immunity (Fig. 2.3).

The pre-index case response

To battle the COVID-19 pandemic within Lagos State and the LGAs, the Lagos State Government developed and published a COVID-19 Response Plan. This plan outlined the mechanisms, initiatives, resources, and timetables that would be necessary for responding to the public health emergency. More broadly, the plan drew on the established capabilities in the State—considering the peculiarities of the threat posed by COVID-19, while also considering applicable lessons learned from previous outbreaks.

When the Lagos State EOC was activated in January 2020, the COVID-19 response strategy was put into action. The first action in this strategy was establishing an Incident Command System (ICS) four weeks before the index case was detected in Nigeria. Under this system, the Lagos State Governor acts as the Incident Commander and is supported by the State's Commissioner for Health. Together, they set the strategic course for the State's response to the COVID-19 pandemic. The ICS was then further divided into six operational pillars: Advocacy and stakeholder management; Research and manuscripts; Incident Manager; Strategy and reporting; Supply

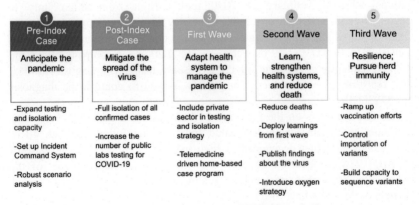

FIGURE 2.3 Summary of the various periods of the COVID-19 pandemic.

chain management and Operations (Fig. 2.4). Domestic and international partners, such as the Nigerian Centres for Disease Control and the WHO, respectively, actively supported the ICS by lending expertise and resources to the pillars, as appropriate.

The ICS went further to create a small but nimble team of executive cabinet members that comprised the war cabinet (WC). The WC sought to balance the economic, social, and health impacts of the pandemic response and implement a tripod pandemic response management strategy. More specifically, this strategy strove to contain the surge of COVID-19 patients while limiting collateral damage to normal health care systems and functions, promoting the ability of citizens to carry out daily economic activities as a means of providing daily subsistence and preventing increases in poverty levels, and maintaining a calm, confident, peaceful, and compliant community. Chaired by the Executive Governor, the WC was empowered to make decisions regarding the changing or modification of policies, as well as decisions regarding the allocation of resources, which were then fast-tracked and made available within 48 hours. These decisions were made based on real data gathered the preceding day to manage the rapidly changing dynamic nature of the pandemic.

The WC met virtually every night and physically one time per week, and press briefings and other public information events were delivered after every physical meeting to update the public on response efforts. These risk communication efforts informed the public on local, national, and international dynamics of the outbreak and the anticipated waves of infection. This approach did well to instill a sense of calmness and confidence, which later

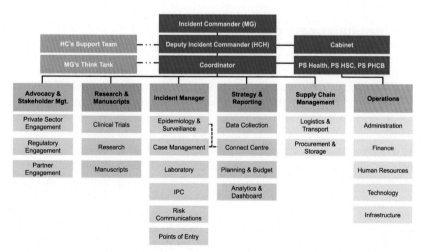

FIGURE 2.4   Structure of the COVID-19 Incident Command System in Lagos State.

resulted in effective civil obedience and complaint citizenry once the pandemic arrived.

A COVID-19 think tank was also established as part of the response to COVID-19. The think tank was a human resource strategy that included a diverse team of subject matter experts from the private sector and academia that analyzed the overall response plan on a regular basis and provided the State with advice on how to enhance the pandemic response. More specifically, this dynamic team includes experts in governance, research, health management information systems, donor engagement, technical support, infection prevention and control, psychological interventions, public relations, and communications that offer direction in various areas, including strategy, clinical governance, communication, logistics, coordination, and project management. Ultimately, the think tank works to promote rapid strategic and policy decision-making for the WC to deliberate on every evening. Decisions made were quickly disseminated to the public, the media, the government, and interest groups participating in the formulation of public policy.

The State also embarked on significant infrastructural and human resources preparedness efforts prior to the arrival of the index case to prepare for the anticipated surge. All existing isolation facilities were fully refurbished and equipped, new facilities were constructed, intensive care units were established, and hotels and hostels were occupied and converted into isolation centers to provide treatment to the thousands of anticipated cases. In collaboration with the Lagos State Asset Management Agency, the Lagos State Ministry of Health was also able to develop the guidelines for Isolation 1 and 2 Facilities, which were subsequently shared with federal government agencies.

The post-index case response

Because the ICS was already in place before the first case of COVID-19 was confirmed in Nigeria, it was easier to mobilize resources and quickly respond once a suspected case arrived. For instance, once the first case of COVID-19 was suspected in Lagos, the case was confirmed within six hours after arriving at the Lagos Mainland Infectious Disease Hospital; this, as opposed to the first case of EVD in 2014, which took three days to diagnose clinically. As briefly discussed earlier, this first case involved a 44-year-old Italian male arriving in Lagos on a Turkish Airlines flight from Milan on February 24, 2020. After spending a night at a hotel close to the airport, the man traveled to his office in Ewekoro, Ogun State, which shares a border with Lagos State to the north. After two days, the man began to feel unwell and sought medical care. In accordance with guidelines from the Nigerian Centres for Disease Control, the EOC was promptly notified, and appropriate response actions were taken.

However, once this index case was confirmed, additional measures were taken that focused on mitigating the spread of the virus. Included in these measures were implementing point-of-entry screenings, social distancing, conducting risk communication sessions, and enhancing surveillance efforts. Point-of-entry screenings involved screening passengers at airports, seaports, and land borders using temperature checks and questions about symptoms of the COVID-19 virus. Infection prevention and control training were also provided for point-of-entry staff and information, education, and communication materials were also distributed to staff, stakeholders, and customers.

Travel restrictions were also announced on March 18, 2020, some three weeks after the first case had been identified. This delay provided ample time for the importation of the virus into the country, as returnees from abroad comprised the majority of confirmed cases early in the pandemic. On March 29, interstate lockdowns were first placed in three states with high incidence, including Lagos. This lockdown was followed by interstate travel restrictions announced on April 23, 2020 for all 36 states and the Federal Capital Territory in the country.

The social distancing strategies implemented aimed to reduce physical contact between people, as a means of reducing the risk and spread of COVID-19 within a community. To enforce this, the federal government of Nigeria prohibited large gatherings, issued compulsory stay-at-home directives to nonessential public servants, and also temporarily closed schools, markets, and places of worship. Individuals were also encouraged to wear face masks and practice good hand and respiratory hygiene practices.

The risk communication sessions that were held sought to spread the word about the risks posed by the virus and the protective measures that could be taken. These were conducted with stakeholders in government agencies and parastatals, the private sector, education, religious institutions, and the unorganized sector. Importantly, risk communication messages underwent a continuous process of refinement. This process was informed by the results of focus group discussions that were held outdoors at motor parks and included 10 representatives from the 376 wards of Lagos. The results produced by these discussions were used to improve the existing COVID-19 risk communication messaging. Other risk communication efforts involved developing leaflets, jingles, and videos with the NCDC that were disseminated using television, radio, and social media to sensitize the public on virus transmission and infection dynamics.

Several training sessions and capacity-building exercises were also conducted. Market men and women, youth organizations, social mobilization committees, ward health committees, head teachers, principals, community-based organizations, artisans, religious organizations, and traditional birth attendants were the target audiences for the training. In total, 125 markets and 25 motor parks located in all 20 LGAs in Lagos hosted market and motor park sensitization activities. Additionally, there were several press appearances on local television networks.

Surveillance efforts were enhanced primarily by providing 400 Disease Surveillance and Notification Officers with detailed training that focused on case definitions for COVID-19, active surveillance, contact tracing, case investigation, and other reporting methods. This training allowed for contacts of confirmed cases to be quickly recognized, located, and monitored for the duration of the incubation period (i.e., 14 days).

Additionally, to scale up diagnostic testing for COVID-19, Lagos State worked to establish 24 LGA walk-in collection locations around the State and conducted a state-wide active case search in May 2020. Through these efforts, samples were collected from community members who met the case description for suspected cases of COVID-19 and sent for laboratory testing. Further, in addition to these broad, population-wide efforts, enhance surveillance efforts were also extended to selected populations. For instance, passengers on aircraft with a confirmed COVID-19 case were actively monitored; similarly, health professionals who were exposed to confirmed COVID-19 were monitored to ensure that they did not become ill and transmit the disease further.

The response to pandemic waves of COVID-19

During the first pandemic wave, the primary goal was to deploy measures to "flatten the curve," or, put another way, reduce the caseload of COVID-19 as a means of reducing the burden of cases below the threshold capacity of the health care system. These measures included an emphasis on reducing infection in the most vulnerable populations while allowing for some degree of controlled community transmission.

Other efforts during the initial pandemic wave included deploying countermeasures while also increasing health system capacities. For instance, 16 IPC standard operating procedures and recommendations were developed and distributed to both key personnel. These materials were accompanied by IPC protocol training for both clinical and nonclinical medical health care staff. Lagos State's public and private health care institutions were also assessed for their readiness to handle COVID-19 cases.

A supply chain strategy was also designed to improve the logistics of the pandemic response. This strategy described the movement of materials and information, storage and distribution plans, human resources, and procedures for last-minute purchases—ultimately leading to the establishment of several key dashboards and systems, such as the Combined Daily Stock Status Dashboard, the Logistics Management Information System, and processes for emergency procurements. These efforts played a crucial role throughout the pandemic response by supporting decision-making regarding the purchase and replenishment of vital commodities.

Additional health system capacity-strengthening efforts focused on bolstering laboratory services. At the onset of the COVID-19 pandemic, Lagos State maintained the laboratory capacity to test up to 180 samples per day. Health officials quickly realized that this capacity would need to be expanded to meet the increasing demand for diagnostic services, so concerted efforts were made to expand laboratory capacity. These included hiring and training more personnel, creating SOPs and job aids, standardizing processes across all laboratories in Lagos State, and supporting laboratory operations to allow laboratories to offer 24-hour services. As a result of these efforts, three months later, the number of samples that could be processed had grown to 2,000 samples per day. A favorable environment for public-private partnerships also helped to expand laboratory capacity and health authorities partnered with private laboratories throughout the state starting in July 2020 to increase the number of diagnostic tests that could be conducted. Recent data show that a total of 700,315 diagnostic COVID-19 tests have been conducted—with public laboratories processing 191,735 samples (27%), and private laboratories processing 508,580 samples (73%).

Relatedly, public-private partnerships also helped to improve the management of COVID-19 cases. Thanks to political support and leadership provided by the government, and financing provided by banks and private organizations several isolation centers were constructed and quickly outfitted. Personal protective equipment was also procured and distributed to medical facilities across Lagos. As a result, the number of isolation centers quickly increased from one to six, and the number of beds at the Lagos Mainland Infectious Disease Hospital from 115 to 590.

Learning from the responses mounted in other regions of the world, health authorities in Lagos knew that the availability of oxygen would be critical for an effective pandemic response. As such, a robust oxygen strategy was introduced to meet the increasing need for oxygen therapy at isolation and treatment centers. This strategy involved outsourcing oxygen supplies at the infectious disease hospital to a third-party logistics firm. This significantly improved the supply of oxygen at the hospital. Additional oxygen therapy centers were subsequently deployed to high-burden LGAs to meet the increased demand for oxygen in those communities. An additional oxygen plant was also deployed at Gbagada General Hospital and an oxygen plant at Lagos University Teaching Hospital was reactivated to increase the local supply of oxygen.

Despite these capacity increases, it soon became clear that not all confirmed cases of COVID-19 could be isolated in health facilities. In response to this challenge, in September 2020, the isolation strategy for COVID-19 cases was adapted to include a home-based care program for mild cases of the disease. This program leveraged a robust telemedicine platform to provide constant, integrated care to those who were confirmed to have COVID-19 but did not require specialized medical care. Care packs—consisting of medical masks, a digital thermometer, zinc tabs, paracetamol tablets, Vitamin C, and Vitamin D—were distributed to confirmed cases of disease. A telemedicine initiative, named Eko Telemed, was also designed and launched to cater to a diverse population of patients—offering services in the four major languages in Lagos State (i.e., English, Yoruba, Hausa, and Igbo).

These efforts were what were primarily used to manage the pandemic waves until COVID-19 vaccines became available and vaccination efforts commenced in March 2021. Lagos developed a robust vaccination strategy that utilized the strengths of both the public and private health care sectors. First and foremost, the strategy sought to reduce deaths by prioritizing the vaccination of those at the highest risk of severe disease and death. The strategy also prioritized vaccinating health care workers as a means of improving health care system resilience. Other strategic priorities included regulating the distribution of vaccine through a central planning system to prevent clinical harm and ensure the best clinical outcomes for all Lagos State residents, and preventing economic harm by targeting those whose absence would significantly impact the economy or COVID-19 response efforts in the short-term (i.e., essential workers).

Lessons learned from Lagos' response to the COVID-19 pandemic

While some of Lagos State's preparedness was built on the knowledge and experiences gained from the 2014 EVD epidemic, many lessons were learned throughout the COVID-19 pandemic. For example, the state has been able to collect intelligence reports daily, identify impending threats to public health, and ensure the coordination of response activities through the use of an incident command center.

The EOC was also formed in Lagos State before the pandemic started, which helped to improve preparedness and launch a rapid response as the virus neared Lagos. For instance, upon laboratory confirmation, treatment for the index patient and contact tracing both began immediately at the infectious disease hospital. In Lagos, several other interventions that appear to have helped to control the pandemic include social distancing, lockdowns, and restrictions on intra- and interstate transportation. Similar response techniques were rapidly used in many other countries with comparable success.

Building health workforce capacity was also a crucial part of the response to COVID-19 in Lagos. Since the outbreak of the pandemic, however, medical personnel have faced a variety of challenges, including exhaustion and burnout caused by overwhelming and unsustainable workloads. This issue must be addressed in the future to ensure that frontline health care workers are capable of properly providing the care that patients need in such a crisis. For instance, some have suggested that future training activities should incorporate coping skills for improving mental health.

Lagos was also able to improve the response and reduce the costs associated with the pandemic response by collaborating with the private sector. For example, private sector actors helped to fund the rapid procurement of urgently needed medical equipment and the construction of isolation centers for the treatment of confirmed cases. Consequently, assistance from the private sector is vital for implementing rapid and robust responses to public health emergencies, such like epidemics and pandemics.

Finally, the WC functioned to balance and deploy the principles of a tripod pandemic response. Maintaining a balance between the three legs of this tripod is a delicate role. For instance, the WC must ensure that the public health response does not overwhelm other considerations and precipitate economic hardship, especially in a population where many individuals live on a daily income. Unsuccessful attempts to balance these considerations could have triggered mass poverty and breakdowns in law and order, with collateral damages far exceeding those caused by the pathogen alone. Experiences from the response to COVID-19 in Lagos suggest that balancing these considerations requires the constant review of real-time data and emerging evidence bases, as well as a deep knowledge of the local context and epidemiological trends.

Conclusion

Pandemic preparedness is a constant process that requires ongoing activity, financing, collaborations, and political commitment on all fronts. The planning, investment, and implementation of key actions depend upon all stakeholders working together. To this end, Lagos State utilized knowledge and lessons learned from the 2014 EVD outbreak to respond to the COVID-19 pandemic. More specifically, there were several existing mechanisms that were adapted and modified to respond to the COVID-19 pandemic.

Strong leadership, partnerships with domestic and international partners, and innovative cooperation with and contributions from the private sector have all benefited the state's response. Still, the experiences from responding to COVID-19 have laid bare that strong health systems are fundamental to effective responses, and the state must continue to improve its resilience and capacity to respond to public health emergencies, like COVID-19. Beyond the immediate benefits to Lagos State, other states and nations could benefit from the experiences, knowledge, and best practices identified and learned through the response to this public health emergency.

References

[1] Lagos State Government. About Lagos. https://lagosstate.gov.ng/about-lagos/. [accessed 07.11.22].

[2] Government of Nigeria – National Bureau of Statistics. National population estimates. https://nigerianstat.gov.ng/elibrary/read/474; 2006–2016 [accessed 04.11.22].

[3] United Nations, Department of Economic and Social Affairs, Population Division. World urbanization prospects: the world's cities in 2018. New York: United Nations; 2018.

[4] Lagos Bureau of Statistics, Ministry of Economic Planning and Budget. Abstract of local government statistics. Ikeja: Lagos State Government; 2019.

Chapter 3

COVID-19 in Beirut: epidemiology, response, gaps, and challenges

Nada M. Melhem, Farouk F. Abou Hassan and Mirna Bou Hamdan
Medical Laboratory Sciences Program, Division of Health Professions, Faculty of Health Sciences, American University of Beirut, Beirut, Lebanon

Background

Lebanon is a small country $(10,452 \text{ km}^2)$ in the Eastern Mediterranean Region overseeing the Mediterranean Sea with an estimated population of 6.7 million (2022) [1]. The youth population (i.e., less than 24 years of age) accounts for almost 44% of the whole population [2]. Lebanon is divided into eight governorates: Akkar, Baalbek-Hermel, Beirut, Bekaa, Mount Lebanon, Nabatieh, North Lebanon, and South Lebanon (Fig. 3.1) [3]. The capital, Beirut, is the largest city in Lebanon with a population of more than 2 million [4]. Beirut is located in the Beirut governorate, which includes the capital city, as well as suburbs, and occupies approximately 2% of Lebanon's total area [5].

Lebanon hosts the largest number of refugees per capita and per square kilometer in the world—with an estimated 1.5 million Syrian refugees, 210,000 Palestinian refugees, and more than 13,700 refugees of other nationalities [6,7]. The majority of refugees live in crowded camps that are characterized by suboptimal living conditions and lack access to safe water and sanitation [8,9]. These living conditions can expose refugees to infectious disease outbreaks and pose additional challenges for governments as they seek to properly respond and contain the spread of outbreaks. This was evident during a recent cholera outbreak in October 2022 whereby the index case of cholera emerged from an informal settlement in North Lebanon [10]. The large number of refugees has been linked to an increased burden on the national economy and to the weakening of the already existing fragile health care system [11].

FIGURE 3.1   A map of Lebanon and the eight administrative governorates; Beirut, the capital of Lebanon, is one of those jurisdictions.

In Lebanon, the first confirmed case of COVID-19 was reported on February 21, 2020 [12]. Since then, more than 1,223,355 cases and 10,750 deaths have been reported as of January 5, 2023 [13]. By mid-March 2020, after the World Health Organization (WHO) characterized the outbreak as a pandemic, the Lebanese government declared a state of emergency and imposed strict public health measures and social measures to reduce the spread of the SARS-CoV-2 virus. These measures included the closure of points-of-entry, travel restrictions and bans, and complete or partial national lockdowns, in addition to mandates requiring the use of face masks, social distancing measures, and quarantine following exposure to a suspected or confirmed COVID-19 case [14,15]. These response measures aimed at mitigating and delaying the steep rise in the number of cases and hospitalization rates and consequently, reducing the pressure on health care systems and health care workers.

These mitigation measures, however, also exacerbated an already fragile socioeconomic situation in the country. Lebanon has been suffering from a severe economic and financial crisis described by the World Bank as the top three most severe crises globally [16]. Moreover, amid a growing pandemic, the Beirut port explosion on August 4, 2020, resulted in more than 220 deaths, 6500 wounded, and left 300,000 people displaced; importantly, the explosion also impacted half of Beirut's health care centers and shifted attention from caring for COVID-19 patients to caring for the victims of the blast [17,18].

This chapter advances the epidemiology of COVID-19 in Lebanon with a specific focus on the city of Beirut and examines gaps and challenges that have been identified throughout the response to the COVID-19 pandemic.

The socioeconomic status in Lebanon

October 2019 marked the beginning of an unprecedented economic and financial crisis in Lebanon [19]. Several months later, the crisis was thereafter exacerbated by the COVID-19 pandemic and the Beirut Port explosion on August 4, 2020. Before the crisis, Lebanon was classified by the World Bank as an upper-middle-income country; however, by July 2022, the World Bank had reclassified Lebanon as a lower-middle-income country [20]. In 2019 the nominal gross domestic product was $52 billion before it plummeted to $23.1 billion in 2021, amounting to a drop of more than 36% between 2019 and 2021 [20]. Political deadlock, corruption, and an inability to implement meaningful economic reforms by Lebanese officials fueled national protests all over the country. Consequently, the Lebanese pound lost more than 90% of its value against the United States dollar, which led to a year-by-year inflation of 120% between May 2020 and May 2021 [19]. Lebanon witnessed a drastic collapse in basic services and an inability to import basic goods including necessary medical devices and essential medications due to the depletion of foreign exchange reserves and the implementation of informal capital controls by the banking sector [20]. The start of the COVID-19 pandemic and the implementation of intermittent national lockdowns exacerbated and deepened the socioeconomic crisis. As a result of the government's inability to address the root causes of the financial crisis, more than 19% of the population reported the loss of their main source of income—with a substantial number of families reporting being unable to afford or access basic goods and services [19].[1] The national unemployment rate reached 29.6% in 2022, compared to an estimated unemployment rate of 11.4% in 2019 [20].

The health care system in Lebanon

The structure of the national health care system in Lebanon had a major impact on the national response to COVID-19. Hospitals and health care facilities with various capabilities are distributed throughout Lebanon, but approximately 84% of the country's health care facilities are located in the big cities—most of them are owned and managed by private sector actors (Fig. 3.2) [12]. Indeed, the number of public hospitals is limited across

---

1. These include safe food and water, health, education, and electricity, among others.

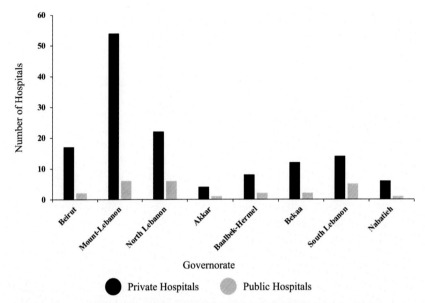

FIGURE 3.2 Distribution of private and public hospitals in Lebanon. *Data for the figure were based on available data from MoPH: https://moph.gov.lb/HealthFacilities/index/3/188/8? facility_type = 8&district = Beirut&name = .*

Lebanon, and when compared to private hospitals, the former are generally underfunded, understaffed, and not well equipped. This is broadly attributable to the allocation of the health care budget; in 2018, only 1.8% of the Ministry of Public Health's (MoPH) budget was invested in public hospitals, whereas more than 80% is invested in private hospitals and pharmaceuticals [21].

The Rafik Hariri University Hospital (RHUH) is located in Beirut and is the largest public hospital in Lebanon. At the beginning of the pandemic, the RHUH was the only hospital nationwide designated for COVID-19 testing as well as admitting, isolating, and caring for confirmed COVID-19 patients [12]. Accordingly, any suspected COVID-19 case outside the capital had to be transferred to RHUH for testing and isolation. As witnessed throughout the COVID-19 pandemic, the limited availability of these services, and limited access to health care in rural areas more broadly, posed significant challenges.

Since the detection of SARS-CoV-2 in Lebanon, the Lebanese government has responded to the pandemic using multiple public health interventions to reduce community transmission; the former included the declaration of a state of national emergency with a complete national lockdown (March—May 2020), the enforcement of traffic regulations based on odd/ even rationing of vehicles (April—June 2020), travel bans (March—July

2020), the implementation of nonpharmaceutical interventions (e.g., social distancing and isolation of confirmed cases), as well as mask mandates which were imposed on May 29, 2020 (Fig. 3.3) [22,23]. These public health and social measures delayed the dramatic surge in the number of COVID-19 cases and deaths; and, importantly, they allowed health care facilities to pre-pare for the isolation and treatment of COVID-19 cases, scale-up surveil-lance and contact tracing efforts, and bolster testing and diagnostic capacities.

The first national lockdown imposed by the government from mid-March until May 2020 allowed the MoPH to equip and prepare public hospitals and health care facilities other than the RHUH across the country to scale up COVID-19 testing and surveillance efforts, admit COVID-19 patients, and implement contact tracing (Fig. 3.3) [12]. These measures alleviated some of the pressure on RHUH for admitting and caring for COVID-19 patients, as well as for COVID-19 testing. Despite the government's efforts to bolster hospital preparedness, several challenges were encountered. These included the poor engagement of the private health care sector during the initial response. More specifically, there was a lack of public-private partnerships, and a tendency of private health care actors to seek for-profit services, rather than opening their facilities to COVID-19 patients or repurposing them to better support the initial response. And, in the context of the low health

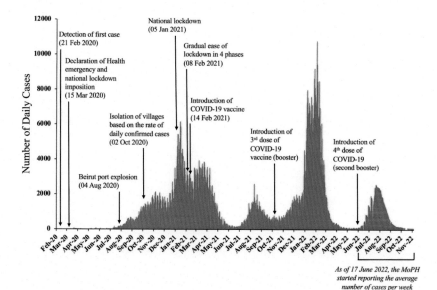

FIGURE 3.3 The national response to the COVID-19 pandemic in Lebanon (February 2020−November 2022). This figure summarizes the different public health and social measures implemented before and after the introduction of COVID-19 vaccines and the corresponding number of daily confirmed COVID-19 cases in Lebanon.

budget assigned to public hospitals, this lack of engagement contributed to the ongoing financial crisis and the heavy physical, mental, and emotional burden felt by health care workers [21,24].

Nevertheless, there were increases in both hospital capacity to treat COVID-19 patients and the supply of personal protective equipment; these were made possible through generous donations and support from various international organizations [25]. The first lockdown also helped several major academic hospitals in Beirut—including the American University of Beirut Medical Center (AUBMC), the Lebanese American University Medical Center-Rizk Hospital (LAUMC-RH), and St. George Hospital—to establish dedicated COVID-19 units to admit and care for COVID-19 patients. These steps were critical as they preceded the first wave of COVID-19 that occurred between December 2020 and February 2021 during which the average number of daily cases reached more than 2745 (Fig. 3.3) with an average positivity rate of 17% (Fig. 3.4). These parameters were more than sixfold and threefold higher than those reported between February and November 2020, respectively. Moreover, on average 749 daily cases needed admission to the intensive care unit (ICU) between December 2020 and February 2021 (Fig. 3.5), with an average daily number of deaths at 37 and a case-fatality rate of 1.5%.

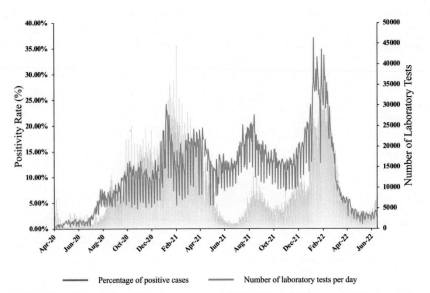

FIGURE 3.4   The number of laboratory tests and COVID-19 positivity rates in Lebanon (April 2020–June 2022). The thicker, orange line represents the number of laboratory tests per day; the thin, blue bars represent COVID-19 test positivity rates. Positivity rates were calculated by dividing the number of confirmed cases by the total number of tests performed per day.

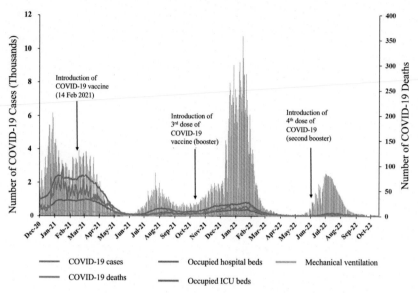

FIGURE 3.5 COVID-19-associated hospitalizations in Lebanon (December 2020–November 2022). The thin, purple bars represent the number of daily COVID-19 cases, and the thick, teal line represents the number of reported COVID-19 deaths. The thick, blue line represents the number of occupied hospital beds; the thick, red line represents the number of occupied ICU beds; and the thick, green line represents the number of patients under mechanical ventilation.

The COVID-19 response in Beirut

The epidemiology of COVID-19 in Beirut

At the beginning of the pandemic and before the port explosion, Beirut averaged five confirmed COVID-19 cases and four reported deaths per day between May and July 2020 (Fig. 3.6).[2] The low morbidity and mortality rates in Beirut were similar to those observed nationwide—with the average number of daily cases in Lebanon amounting to 42 cases and a total of 35 deaths reported during the same period.

Toward the end of April 2020, a five-phase plan for easing the national lockdown was put in place. This strategy sought to gradually lift the lockdown measures while monitoring the impacts on community transmission levels. The first phase started on April 27, 2020, and included the opening of vital economic sectors that posed a low risk for viral transmission. Two weeks later, the lockdown was further eased, whereby vital economic sectors with a higher risk of transmission and businesses with lower economic effects were reopened in the second and third phases, respectively [12]. The

---

2. It is not possible to calculate positivity rates in Beirut due to the lack of information on the number of COVID-19 laboratory tests performed in the city.

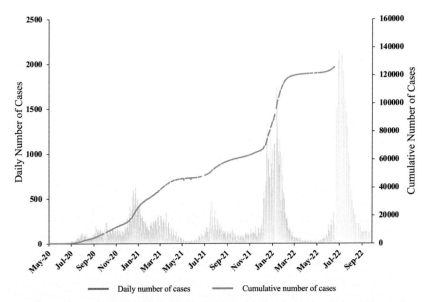

FIGURE 3.6   The burden of COVID-19 in Beirut (May 2020−October 2022). The thin, blue bars represent the number of confirmed cases per day and the thick, orange line represents the cumulative number of cases across time. As of 15 June 2022, the MoPH started to report the number of COVID-19 cases in Beirut in the past 7 days.

fourth phase targeted the reopening of the retail and recreational centers on June 8, 2020, and the final phase targeted the lift of international travel bans on July 1, 2020, and the opening of the airport [26].

However, following the Beirut port explosion in August 2020, the average number of daily cases in Beirut increased by more than 13-fold and the total number of deaths caused by COVID-19 reported in August 2020 was 29; the latter is over seven times higher than the total number of deaths reported in the city between May and July 2020. The 14-day average incidence rate per 100,000 in August 2020 was 198. Nationally, the average number of daily cases in August 2020 was 407 with an average daily positivity rate of 6% and a cumulative number of deaths of 108 during that month.

This dramatic surge in the number of cases and deaths following the blast was due to shifting attention away from adhering to nonpharmaceutical intervention measures toward saving lives and caring for casualties resulting from the blast [18]. This large number of casualties that demanded immediate medical care and attention resulted in crowded emergency departments in hospitals across Beirut, often exceeding the capacities of health care facilities. Moreover, the blast heavily damaged four hospitals in Beirut rendering them nonfunctional—with one of these being St. George Hospital, located less than 1 km from the Port of Beirut; it is also one of the largest hospitals in Beirut and was one of the facilities caring for COVID-19 patients [27].

This situation resulted in a reduction in both the number of regular hospital beds, as well as ICU beds required for COVID-19 patients. The number of daily COVID-19 cases in Beirut kept on gradually increasing with a daily average of 118 cases between September and December 2020 and a 14-day average incidence rate of 322 per 100,000 (Fig. 3.6). Similarly, the average daily number of COVID-19 cases nationwide during that period was 1351, with a 14-day average incidence rate of 374 (Fig. 3.3). During that time, community transmission—based on criteria established by the WHO, whereby four distinct levels were defined as low, moderate, high, and very high incidence of locally-acquired, widely dispersed cases detected in the past 14 days—reached its highest level (i.e., Level 4). These levels accounted for the incidence rate, deaths, and hospitalizations, as well as testing positivity rates [28]. In response to this and to reduce the impacts on the already fragile health care system, a 2-week lockdown was enforced between August 21 and September 3, 2020, with 5 days of full lockdown, followed by 2 days of partial lockdown, repeatedly [29]. These measures, however, failed to reduce the community spread of COVID-19 possibly due to lack of compliance. In September 2020, the average number of daily cases in Beirut was 78, with a total of 31 deaths reported that month; in October 2020, the average number of daily cases in Beirut increased by 1.5 fold, with a total of 20 deaths.

The continuous increase in the number of COVID-19 cases along with the lack of compliance with public health and social measures following the port explosion motivated the introduction of a new approach geared toward controlling the spread of the virus in the community. In October 2020, the government began a "green zoning" approach, whereby villages were isolated based on the average rate of daily cases per 100,000 residents over the previous 14 days [30]. The risk level for community transmission was divided into three tiers—low (less than 4 cases), moderate (4−8 cases), and high (greater than 8 cases) [31]. Consequently, by mid-October, a complete lockdown of more than 160 villages was implemented. This lockdown was supported by engaging a large number of municipalities and was accompanied by the closure of pubs, bars, and night clubs, and the enforcement of a nationwide, mandatory indoor face mask policy [32]. However, the implementation of this approach did not dramatically change the case fatality rate, nor did it slow the level of community transmission as the 14-day average incidence rate increased from 206 to 433 per 100,000 population between September 25, 2020, and November 14, 2020. Consequently, the government imposed an additional national lockdown on November 14, 2020, prior to the holiday season.

However, this lockdown was short-lived. In an attempt to boost the economy, the Lebanese government decided to ease lockdown measures two weeks later and open entertainment venues during the 2020−2021 holiday season, most of which are located in the Greater Beirut area. The easing of these measures coincided with the detection of the first SARS-CoV-2 variant of concern, Alpha, and a sharp increase in the number of COVID-19 cases, the incidence rate, and COVID-19-related deaths in Beirut [33]. In January 2021, the average number of daily reported cases was 361, with a 14-day average incidence rate of 1013 per 100,000, and a total of 124 reported deaths (Fig. 3.6).[3] The epidemiology of COVID-19 in Beirut during this time was a reflection of the pandemic nationwide whereby the average daily number of reported cases surpassed 3800, with an average daily positivity rate of 20% and an incidence rate of 867 per 100,000 people (Fig. 3.4). These increases in morbidity were coupled with dramatic increases in ICU occupancy rates, which reached greater than 90% nationwide [33], and a sharp increase in reported mortality, which prompted the government to impose another nationwide lockdown on January 15, 2021, that extended until February 8, 2021 (Fig. 3.3).

COVID-19 vaccination: successes and challenges

The introduction of COVID-19 vaccines in Lebanon was arguably the most important success in the pandemic response. On February 14, 2021, the national COVID-19 vaccine plan was launched following the delivery of the first batch of the Pfizer-BioNTech vaccine [34]. The IMPACT online platform was launched to support vaccine administration, aid in vaccine registration, and monitor adverse events associated with vaccination. The national COVID-19 vaccine plan initially targeted individuals based on their risk of exposure and infection, their risk of complications following infection, and their status as essential personnel (e.g., front-line responders to the pandemic, health care workers, etc.), while considering the availability of vaccine doses [34]. These priorities and considerations for the administration of COVID-19 vaccines were aligned with recommendations made by the WHO Strategic Advisory Group of Experts Framework for the allocation and prioritization of vaccination in targeting high-risk groups at different stages of vaccine supply availability [35].

Following the start of the COVID-19 vaccination campaign, the number of COVID-19 cases, hospitalizations, and deaths nationwide began to gradually decrease over time. Nationwide, from May 1, 2021, to July 31, 2021, there was an average of 378 cases per day, with an average daily positivity rate of 2.3% nationwide (Figs. 3.3 and 3.4). The total number of deaths during that time was 628, with an average of seven daily deaths, and a case-fatality rate of 2.4%. Similar trends were observed in Beirut, with low average daily cases (52 cases), and a total of 53 deaths (Fig. 3.6). Declines were also observed in the average number of patients requiring hospitalization per day and the number of patients requiring ICU admission and mechanical

---

3. Data on hospitalization rates in Beirut during this time were, unfortunately, unavailable.

ventilation—83% and 78%, respectively—compared to January 2021 (Fig. 3.5). While there are no data available on hospitalization rates in Beirut, the reduction in the latter observed nationwide reflect trends in hospitalizations in Beirut. While the decline in COVID-19 cases over this period may largely be attributed to COVID-19 vaccines, other factors likely contributed to the decline—including the continuation of remote schooling and working from home, the increased use of masks, social distancing practices, and natural immunity acquired following infection.[4]

At the time of writing this chapter, the COVID-19 vaccine in Lebanon was available in vaccination centers across the country to everyone 12 years of age or older on a walk-in basis [36]; with eight of these centers located in the Beirut governorate and 47 distributed across other governorates [37]. Scaling-up vaccination across all age groups in Lebanon was instrumental to speed up vaccination rates and coverage. Recent reports from the MoPH on COVID-19 vaccine coverage nationwide show that 50.2% of the population has received a first dose of COVID-19 vaccine, and 44.2% have received a second dose [38]; further, of those who have received a second dose, 27.3% have received a third dose (i.e., booster dose) [38]. Still, these figures lag far beyond the targets set by the national COVID-19 vaccine plan, which aspired to vaccinate at least 70% of the population by the end of 2022. Moreover, because of aggregated data reporting, there is limited information on vaccine coverage in subnational jurisdictions, rendering it difficult to monitor coverage in specific locations, such as Beirut, and coverage in specific subpopulations.

The emergence and monitoring of variants of concern

In August 2021, a new surge in COVID-19 cases was observed in Beirut with average daily case counts rising to 205, and an average 14-day incidence rate of 555 per 100,000 people (Fig. 3.6). This wave was primarily driven by the emergence and subsequent circulation of the Delta variant—a highly transmissible variant of concern associated with more severe disease compared to previously circulating variants [39]. National trends in Lebanon largely mirrored global ones, whereby the average number of daily cases in Lebanon peaked at 2085 cases per day in September (Fig. 3.3). Similar patterns were also observed in terms of average daily positivity rate, which rose to 5.3%, and incidence rates that climbed to 282 per 100,000 people.

Between October and November 2021, the number of cases gradually decreased before it spiked again between December 2021 and January 2022. Similar to the first wave of COVID-19 in Lebanon, this wave was attributable to the mass gatherings that occurred during the holiday season

---

4. The durability of naturally acquired immunity, however, remains poorly understood at the time of writing.

and New Year's Eve; however, it was also the result of the emergence of a new, highly transmissible SARS-CoV-2 variant of concern—the Omicron variant. During this period, the average number of daily cases in Beirut increased from 174 in December 2021 to 860 in January 2022 (Fig. 3.6). Similarly, the average 14-day incidence rate per 100,000 people in Beirut jumped from 355 in December 2021 to 2093 in January 2022. As was the case with the wave attributable to the Delta variant, these trends reflected those at the national scale, whereby the average number of daily cases increased from 1848 in December 2021 to 6235 in January 2022 (Fig. 3.3). The national average 14-day incidence rate and the average daily positivity rate also jumped from 411 cases per 100,000 people and 8.4% in December 2021 to 1381 cases per 100,000 people and 18.4% in January 2022. Despite these substantial increases, the country did not observe increases in hospitalization rates, which can be attributed to the increased vaccine coverage during that time (47% first dose and 39% second dose) along with natural immunity [13]. This was supported by 81% of patients in ICUs or those requiring mechanical ventilation being unvaccinated—highlighting the ability of vaccines to prevent severe illness, and underscoring the value of COVID-19 vaccines, as well as the importance of scaling-up vaccination coverage across all eligible age groups.

SARS-CoV-2 genomic surveillance is important to continuously monitor the circulation and emergence of variants in order to guide public health policies. Since its emergence, the Omicron variant evolved into multiple descendent lineages [39,40]. These lineages differ by the number of mutations in the spike protein, which is thought to confer higher transmissibility and immune escape characteristics. In response to this, and in collaboration with the MoPH and with support from the WHO, there have been efforts to establish national SARS-CoV-2 genomic surveillance for health care workers and hospitalized patients in Lebanon [41]. Initial data showed that between December 2021 and January 2022, among 250 sequenced samples from health care workers in Lebanon, 19% were Omicron sublineage BA.1, and 57% of the samples were Omicron sublineage BA.1.1 [41]. A majority of the collected samples ($n = 205$) came from three hospitals in Beirut. These results are in accordance with global data whereby 78% and 16% of sequences submitted to GISAID during that time were BA.1 and BA.2, respectively [42]. Several months later, during the early summer, BA.2.12.1, BA.4, and BA.5 sublineages, derived from BA.2, began to dominate the epidemiological trends worldwide [43]. BA.5 sublineage, which first emerged in South Africa, has a growth advantage over the other sublineages and later became the predominant circulating sublineage detected in more than 100 countries [44,45]; and as of October 1, 2022, 82% of the submitted sequences to GISAID were BA.5 [42]. The spread of this highly transmissible subvariant resulted in a fifth COVID-19 wave in Lebanon between June and September 2022 (Fig. 3.3). Nevertheless, hospitalization rates across the country remained low (Fig. 3.5). Moreover, since June 14, 2022, the MoPH stopped reporting the number of COVID-19 laboratory tests performed each day—reflecting a decrease in laboratory diagnostic testing largely due to the availability of rapid diagnostic testing; thus, an underestimation of community spread.

Conclusion: successes, gaps and challenges, and the way forward

The response to the COVID-19 pandemic in Lebanon faced a wide range of challenges. Initially, the RHUH was the only public hospital designated for testing, quarantining, and treating confirmed cases [12]. This resulted in underestimating COVID-19 transmission and the inability to predict trends at the beginning of the pandemic—both of which hampered efforts to make evidence-informed decision-making. On April 3 2020, this gap was addressed as other hospitals—including three major academic hospitals in Beirut—established COVID-19 units and became designated testing centers for COVID-19 [46]. Still, the COVID-19 pandemic highlighted the lack of a comprehensive emergency preparedness plan for a timely response to a public health emergency, such as the emergence and spread of SARS-CoV-2.

Importantly, the banking sector in Lebanon imposed limits on US dollar withdrawals, which subsequently increased prices and devaluated the Lebanese pound against US dollars. Consequently, there have been shortages in medical supplies and equipment, as suppliers have been unable to import these materials due to the shortage of foreign currency [47]. This complicated the pandemic response and posed enormous pressures on health care centers that lacked the necessary amounts of hospital beds, ICU beds, ventilators, and other medical equipment and medications dedicated to treating and managing COVID-19 cases. Moreover, national lockdown measures implemented by the government early during the pandemic—including borders and airport closures—worsened the country's socioeconomic status, resulted in many businesses closing, and cost thousands of people their jobs [48]. All of these challenges have resulted in considerable amounts of emigration and brain drain from Lebanon—especially in the health care sector, where, as of September 19, 2021, an estimated 40% of skilled medical doctors and 30% of registered nurses have left the country [49].

At the beginning of the pandemic, there were also limited capacities for diagnostic testing, contact tracing, and isolating confirmed COVID-19 cases. These were largely due to delayed testing scale-up and diagnostic capacities, an under-resourced and under-staffed surveillance system, insufficient genomic surveillance programs, a lack of data sharing and established decentralized health informatics, lack of digital contact tracing technologies, and shortages in isolation facilities. The dearth of genomic surveillance programs was particularly critical, as these programs are necessary for not only monitoring the circulation and emergence of SARS-CoV-2 variants but also guiding and informing public health policies. While some genomic analyses were performed early during the pandemic, data were sporadic and often characterized by small sample sizes that were not representative of different areas in Lebanon [50,51]. This underscores the vital importance of the previously discussed efforts to conduct national SARS-CoV-2 genomic surveillance for health care workers and hospitalized patients in Lebanon [41].

Despite all these challenges, public health measures imposed by the Lebanese government, in addition to community mobilization, were key factors in delaying the exponential rise in the number of COVID-19 cases in Lebanon. Community mobilization involved the engagement of municipalities that started in the Greater Beirut area through sharing information with the MoPH as well as monitoring home isolation for confirmed and suspected cases under their jurisdiction. The concerted efforts and collaboration between the MoPH, the Ministry of Interior, the Lebanese Red Cross, various United Nations agencies, the WHO, and the Disaster Risk Management Unit resulted in better implementation of the national response against the spread of COVID-19. Volunteers were trained for contact tracing nationwide and were also involved in testing and data collection, specifically at points of entry in airports and at land border crossings. The ability and the speed by which major academic hospitals in Beirut (e.g., RHUH, AUBMC, LAUMC-RH, St. George Hospital, etc.) were prepared to care for COVID-19 patients early during the pandemic, the early compliance of the Lebanese community to the implemented national strategies and safety measures, and the deployment of the COVID-19 national vaccine plan contributed significantly to the success of the pandemic response.

Ultimately, the COVID-19 pandemic has exposed several notable gaps and challenges pertinent to improving public health emergency preparedness, response, and recovery. While an emergency preparedness plan exists in Lebanon, there is a dire need to reimagine preparedness in the country and to create a national strategy that cuts across emergency preparedness, response, and recovery from crises. This strategy should be applied and sustainable for any type of threat including agents of potential epidemic threats as well as biological, chemical, radiologic, and natural disasters and must give due consideration to subnational implementation at the local level and in communities. Other efforts could also include establishing a well-resourced emergency preparedness and response unit at the MoPH, ideally one compliant with the International Health Regulations. This would involve establishing an intersectoral emergency task force to generate and update emergency-related protocols and integrating a One Health approach into national health strategies. Importantly, this strategy should aim to ensure that a plan of action is developed and prioritized that supports the national health sector strategy on health security, and the strengthening of existing infrastructure. These efforts will require reconceptualizing the health care system and its components, with emphases on preventive care, infection prevention and control, national stockpiles (e.g., medicines, vaccines, diagnostics, personal protective equipment, and others), and supply chains. It will also be critically important to strengthen and build surveillance capacities through the strengthening and development of interoperable laboratory networks, genomic surveillance systems, and robust health information systems for transparent data sharing. These systems should have the ability to link epidemiologic, clinical, and genomic data. A commitment to these important components—ones that are inherent to a rapid and effective pandemic response—will require substantial amounts of resources, as well as investments in human and technical expertise to reduce inequalities in access to information and care.

The COVID-19 pandemic revealed the vulnerability as well as the resilience of the Lebanese health infrastructure. The COVID-19 pandemic highlighted the importance of reimagining the health care infrastructure while strengthening the public sector, investing in a decentralized health information system, bolstering surveillance systems and laboratory networks, as well as engaging the community in public health interventions. As a result, efforts are in place to develop a new national strategy for public health emergencies. However, the political and financial crises, the insufficient efforts to implement meaningful reforms by Lebanese officials, and the deteriorating socioeconomic status may render the commitment to these ideas performative and the implementation of these strategies farfetched.

References

[1] United Nations Population Fund (UNFPA). World population dashboard Lebanon, <https://www.unfpa.org/data/world-population/LB>; 2022 [accessed 07.11.22].

[2] International Labour Organization, UNICEF. Synthesis of the crisis impact on the Lebanese labour market and potential business, employment and training opportunities. Geneva: International Labour Organization; 2022.

[3] WorldAtlas. Maps of Lebanon, <https://www.worldatlas.com/maps/lebanon>; 2022 [accessed 07.11.22].

[4] World Population Review. Lebanon Population, 2022, <https://worldpopulationreview.com/countries/lebanon-population>; 2022 [accessed 07.11.22].

[5] Faour G, Mhawej M. Mapping urban transitions in the Greater Beirut Area using different space platforms. Land 2014;3(3):941−56. Available from: https://doi.org/10.3390/land3030941.

[6] United Nations High Commissioner for Refugees. UNHCR Lebanon at a glance, <https://www.unhcr.org/lb/at-a-glance>; 2022 [07.11.22].

[7] United Nations Relief and Works Agency for Palestine Refugees in the Near East. Hitting rock bottom - Palestine refugees in Lebanon risk their lives in search of dignity [EN/AR], <https://reliefweb.int/report/lebanon/hitting-rock-bottom-palestine-refugees-lebanon-risk-their-lives-search-dignity-enar>; 2022 [accessed 14.11.22].

[8] Salti N., Chaaban J., Moussa W., Srour I., Al Mokdad R., Turkmani N., et al. Assessing shelter and WASH conditions of Syrian refugees in Lebanon in relation to cash assistance

and services. Oslo: Cash Monitoring, Evaluation, Accountability and Learning Organization- al Network (CAMEALEON); 2022.

[9]   Action Against Hunger. Lebanon: COVID-19 exacerbates already-impossible living situation for Syrian refugees. Washington, DC: AAH; 2020.

[10]   World Health Organization. Cholera - Lebanon. <https://www.who.int/emergencies/disease-outbreak-news/item/2022-DON416>; 2022 [accessed 15.11.22].

[11]   Brun C, Fakih A, Shuayb M, Hammoud M. The economic impact of the Syrian refugee crisis in Lebanon: what it means for current policies. Ontario: World Refugee & Migration Council; 2021; 2021.

[12]   Khoury P, Azar E, Hitti E. COVID-19 response in Lebanon: current experience and challenges in a low-resource setting. JAMA. the Journal of the American Medical Association 2020;324(6):548−9. Available from: https://doi.org/10.1001/jama.2020.12695.

[13]   Ministry Of Public Health (MOPH). Monitoring of COVID-19 infection In Lebanon. Beirut: MoPH; 2022.

[14]   Al-Amer R, Maneze D, Everett B, Montayre J, Villarosa AR, Dwekat E, et al. COVID-19 vaccination intention in the first year of the pandemic: a systematic review. Journal of Clinical Nursing 2022;31(1−2):62−86. Available from: https://doi.org/10.1111/jocn.15951.

[15]   Mehta V. The new proxemics: COVID-19, social distancing, and sociable space. Journal of Urban Design 2020;25(6):669−74. Available from: https://doi.org/10.1080/13574809.2020.1785283.

[16]   World Bank. Lebanon sinking into one of the most severe global crises episodes, amidst deliberate inaction, <https://www.worldbank.org/en/news/press-release/2021/05/01/lebanon-sinking-into-one-of-the-most-severe-global-crises-episodes>; 2021 [accessed 07.11.22].

[17]   International Federation of Red Cross And Red Crescent Societies. Lebanon/MENA: Beirut-Port Explosion, operation update (August 2020 − July 2021), Appeal No: MDRLB009. Geneva: IFRC; 2022.

[18]   International Medical Corps. Beirut explosion situation report #9, February 10, 2021. Washington, DC: International Medical Corps; 2021.

[19]   ACAPS. Emergency response plan Lebanon -2022, <https://www.acaps.org/country/lebanon/crisis/soci2021oeconomic-crisis>; 2021 [accessed 07.11.22].

[20]   World Bank. The World Bank in Lebanon − Overview, <https://www.worldbank.org/en/country/lebanon/overview#1>; 2021 [Accessed 27 Dec 2022].

[21]   Knowledge to Policy (K2P) Center. Informing readiness and response to COVID-19 in hospitals and primary health care centers. Beirut: American University of Beirut; 2020.

[22]   Republic of Lebanon - Ministry of Information. Cars in Lebanon Will Now Be Allowed on the Roads Based on Odd/Even License Plates, <https://www.ministryinfo.gov.lb/en/46927>; 2020 [Accessed 27 Jun 2022].

[23]   LBCI. Face masks obligatory as of May 29, <https://www.lbcgroup.tv/news/d/breaking-news/523121/face-masks-obligatory-as-of-may-29/en>; 2020 [Accessed 27 Jun 2022].

[24]   Sonnenfeld D. Amid Corona Crisis, Private Hospitals in Lebanon not Doing Enough, Experts Say". The Media Line 2021;31 Jan.

[25]   United Nations Children's Fund (UNICEF). UNICEF is providing supplies and technical support to fight COVID-19 in Lebanon. Available from: https://www.unicef.org/mena/press-releases/unicef-providing-supplies-and-technical-support-fight-covid-19-lebanon; 2020.

[26]   Haddad N, Clapham HE, Abou Naja H, Saleh M, Farah Z, Ghosn N, et al. Calculating the serial interval of SARS-CoV-2 in Lebanon using 2020 contact-tracing data. BMC

Infectious Diseases 2021;21(1):1053. Available from: https://doi.org/10.1186/s12879-021-06761-w.

[27] Meyers T. "The Beirut Blast Left Lebanon's Health System Badly Shaken. Direct Relief 2020;14 Oct.

[28] World Health Organization. Considerations for implementing and adjusting public health and social measures in the context of COVID-19. Geneva: WHO; 2020.

[29] Crisis 24. Lebanon: Nationwide COVID-19 lockdown introduced until September 3 /update 29. <https://crisis24.garda.com/alerts/2020/08/lebanon-nationwide-covid-19-lock-down-introduced-until-september-3-update-29>; 2020 [Accessed 27 Dec 2022].

[30] Ministry of Public Health. Epidemiological surveillance program of COVID-19, <https://www.moph.gov.lb/en/Pages/2/24870/novel-coronavirus-2019->; 2020 [Accessed 10 Jun 2021].

[31] Presidency of the Council of Ministers of Lebanon. Lebanon National Operations Room Daily Report on COVID-19, Report #198. Beirut: Presidency of the Council of Ministers of Lebanon; 2020.

[32] United Nations Office for the Coordination of Humanitarian Affairs. COVID-19 response – Lebanon bi-monthly situation report (16 October 2020)Available from: https://relief-web.int/report/lebanon/covid-19-response-lebanon-bi-monthly-situation-report-16-october-2020. [accessed 27.06.22].

[33] Koweyes J, Salloum T, Haidar S, Merhi G, Tokajian S. COVID-19 pandemic in Lebanon: one year later, what have we learnt? mSystems 2021;6(2). Available from: https://doi.org/10.1128/mSystems.00351-21 e00351-21.

[34] Ministry of Public Health. COVID-19 National Vaccine Plan. Beirut: MoPH; 2020.

[35] World Health Organization. WHO SAGE values framework for the allocation and prioritization of COVID-19 vaccination. Geneva: WHO; 2020.

[36] Ministry of Public Health. COVID-19 National Vaccination Campaign FAQs about COVID-19 Vaccine. Beirut: MoPH; 2022.

[37] Ministry of Public Health. Pfizer Vaccination Centers, <https://www.moph.gov.lb/user-files/files/Prevention/COVID-19%20Vaccine/Pfizer%20vaccination%20centers-en.pdf.>; 2022 [Accessed 8 Nov 2022].

[38] Ministry of Public Health. Monitoring COVID-19 in Lebanon - October 21, 2022. Beirut: MoPH; 2022.

[39] Hirabara SM, Serdan TDA, Gorjao R, Masi LN, Pithon-Curi TC, Covas DT, et al. SARS-COV-2 Variants: Differences and Potential of Immune Evasion. Frontiers in Cellular and Infection Microbiology 2022;11:781429. Available from: https://doi.org/10.3389/fcimb.2021.781429.

[40] World Health Organization. Tracking SARS-CoV-2 variants, <https://www.who.int/en/activities/tracking-SARS-CoV-2-variants/>; 2022 [Accessed 22 Sep 2022].

[41] Al Kalamouni H, Hassan FFA, Hamdan MB, Page AJ, Lott M, Ghosn N, et al. Genomic Surveillance of SARS CoV2 in COVID-19 vaccinated health care workers in Lebanon. BMC Medical Genomics 2023;16(1):14. Available from: https://doi.org/10.1186/s12920-023-01443-9.

[42] GISAID. Genomic epidemiology of SARS-CoV-2 with subsampling focused globally over the past 6 months, <https://nextstrain.org/ncov/gisaid/global/6m>; 2022 [Accessed 08 Nov 2022]

[43] Aggarwal A, Akerman A, Milogiannakis V, Silva MR, Walker G, Stella AO, et al. SARS-CoV-2 Omicron BA.5: Evolving tropism and evasion of potent humoral responses and

resistance to clinical immunotherapeutics relative to viral variants of concern. eBioMedicine 2022;84:104270. Available from: https://doi.org/10.1016/j.ebiom.2022.104270.

[44] Callaway E. What Omicron's BA. 4 and BA. 5 variants mean for the pandemic. Nature 2022;606:848−9. Available from: https://doi.org/10.1038/d41586-022-01730-y.

[45] World Health Organization. Weekly epidemiological update on COVID-19 - 20 July 2022. Available from: https://www.who.int/publications/m/item/weekly-epidemiological-update-on-covid-19---20-july-2022. [accessed 08.08.22].

[46] Ministry of Public Health. List of Hospitals Eligible for SARS-CoV-2 Testing Using RT-PCR. Available from: https://www.moph.gov.lb/en/Media/view/27426/coronavirus-disease-health-strategic-preparedness-and-response-plan-#/en/Media/view/29673/hospitals-rt-pcr-. ; 2020.

[47] Devi Sharmila. Economic crisis hits Lebanese health care. Lancet 2020;395(10224):548. Available from: https://doi.org/10.1016/S0140-6736(20)30407-4 32087781.

[48] Fares Mohamad Y, Musharrafieh Umayya, Bizri Abdul Rahman. The impact of the Beirut blast on the COVID-19 situation in Lebanon. Z Gesundh Wiss 2023;31(4):575−81. Available from: https://doi.org/10.1007/s10389-021-01562-6 34055571.

[49] Ghebreyesus TA, Al Mandhari A. Joint statement by Dr Tedros Adhanom Ghebreyesus, WHO Director General, and Dr Ahmed Al Mandhari, Regional Director for the Eastern Mediterranean, on Lebanon. World Health Organization Regional Office for the Eastern Mediterranean 2021;19 Sep.

[50] Feghali Rita, Merhi Georgi, Kwasiborski Aurelia, Hourdel Veronique, Ghosn Nada, Tokajian Sima. Genomic characterization and phylogenetic analysis of the first SARS-CoV-2 variants introduced in Lebanon. PeerJ 2021;9:e11015. Available from: https://doi.org/10.7717/peerj.11015 34611501.

[51] Fayad N, Abi Habib W, Kandeil A, El-Shesheny R, Kamel MN, Mourad Y, et al. SARS-CoV-2 variants in Lebanon: evolution and current situation. Biology 2021;10(6):531. Available from: https://doi.org/10.3390/biology10060531.

Chapter 4

Pandemic governance and community mobilization in conflict: a case study of Idlib, Syria

Abdulkarim Ekzayez, Munzer Alkhalil, Preeti Patel and Gemma Bowsher
Research for Health Systems Strengthening Syria (R4HSSS), National Institute for Health Research, Centre for Conflict and Health Research, King's College London, London, United Kingdom

Funding statement

AE, MA, and PP are funded through the National Institute for Health Research (NIHR) 131207, Research for Health Systems Strengthening in northern Syria (R4HSSS), using UK aid from the UK Government to support global health research. The views expressed in this publication are those of the author(s) and do not necessarily reflect those of the NIHR or the UK government.

Background

The city of Idlib is the capital of the Idlib Governorate in Northwest Syria, located about 60 km north of Aleppo. The region is the last remaining opposition-held territory in Syria following the onset of conflict over 11 years ago [1]. The Assad regime, supported by Russia, has carried out sustained aerial bombardment in efforts to recapture the territory, despite repeated attempts at brokering a ceasefire by the international community. Since January 2018, and despite Idlib being subject to a de-escalation agreement, attacks have continued to intensify. Now, around a quarter of the remaining Syrian population lives within the Idlib region, which remains under assault by Syrian, Russian, and Iranian coalition forces.

Idlib's status as a center of culture, history, and political identity sets the stage for its deteriorating fortunes following the onset of conflict. Situated halfway between Aleppo and Latakia, the Idlib region was home to several holiday resorts for urban residents across Northwest Syria to escape the heat of summer in the mountains. Hot summers and cold winters produced fertile territory beyond the city. Idlib has long been celebrated for its olive tree and olive oil production, which formed an important element of a vibrant agricultural and trade culture. Proximity to the Turkish border and the connection of Idlib by highway to Aleppo and onwards to Damascus made the city an important trade center connecting Syria's major municipal centers with the agricultural heartlands. Political sentiment in Idlib has long maintained opposition to the Assad regime's authoritarian rule. Demonstrations against the incumbent leader's father, President Hafez al-Assad during his one and only visit to Idlib famously saw him being pelted with tomatoes and a shoe [2]. The city and wider region were subsequently marginalized by the first Assad regime, which was manifested by exclusion from infrastructure development resulting in a protracted period of fermenting antipathy toward the Assad family and its governments.

After the start of the Syrian revolution in 2011, Idlib city was among the first cities to host mass demonstrations against the Regime of Bashar al-Assad. Protests transitioned into armed conflict following violence launched by the Government of Syria (GoS) [3]. Large parts of Idlib City and the surrounding countryside fell under opposition control. Consequently, the city was subjected to large-scale conflict hostilities with rapid changes in control between the GoS and rebel groups. In May 2015, armed opposition factions took over Idlib city prompting the GoS to withdraw entirely from the governorate [4].

Mass population displacement has resulted in what Mark Lowcock, the United Nations (UN) emergency relief coordinator has described as "the biggest humanitarian horror story of the 21st Century" [5]. More than 900,000 people have been displaced into and within the region and around 1.7 million people currently live in approximately 1400 camps for internally displaced persons (IDPs), of which around 500,000 are children [6]. These camps regularly experience water and food shortages, and only 40% of IDPs have access to working latrine facilities [7]. These shortages are a calculated strategy of the Assad Regime,[1] which has cut off electricity, water, and aid supplies to the Idlib region. Presently, around 75% of the population of Northwest Syria is dependent on UN-led aid, 85% of which comes through the Bab al-Hawa crossing from Turkey following the forced closure of the

---

1. Unless otherwise specified, references to the "Assad Regime" refer to the Government led by the Bashar Al-Assad, the incumbent President since July 2000. The "First Assad Regime" refers to rule of Hafez Al-Assad, the Father of the current President who ruled Syria from 1971–2000. The terms "Assad Regime," and the "Government of Syria" (GoS) are synonymous.

three other points of entry via Iraq, Jordan, and Turkey [8].[2] With only this single crossing remaining, the region has been starved of critical health and subsistence resources for a growing IDP and resident population.

Since the beginning of the conflict, the GoS has directly targeted health facilities throughout Northwest Syria with a sustained campaign of violence. The Assad regime has undertaken a deliberate and targeted assault on health care facilities, in contravention of international humanitarian law [9]. The bombing of hospitals, hijacking of aid convoys, and the kidnaping, torture, and murder of health workers have all been featured as part of this enduring assault on the health sector by pro-GoS forces. Across Syria, only 64% of hospitals and 52% of primary health care centers are still functioning, and around 70% of the health care workforce has fled the country [10]. The WHO has reported that "of all armed conflicts across the globe, Syria has for years been one of the worst examples of violence affecting health care" [12]. A renewal of airstrikes between late-2019 and early-2020 was described by the UN as a "bloodbath" [13]. During this period, attacks on health care increased fivefold with over 40 health facilities bombed in opposition-held Northwest Syria. These strikes included the bombing of Idlib Central Hospital in February 2020 leading to a shutdown of the facility [14]. In March 2020, it was reported that were only 166 doctors and 64 health facilities, mostly operating with minimum-capacity infrastructure in Idlib [15].

Both the city and the wider region of Idlib have been subjected to further events constituting assaults on international humanitarian law. In particular, the repeated use of chemical weapons on civilian populations has escalated an already tense security environment [16]. Allegations emerged in 2015 regarding the use of chlorine against civilians, an allegation that was investigated and affirmed by the Organisation for the Prohibition of Chemical Weapons [17]. A further attack using Sarin took place in 2017 in Khan Shaykoun, a town within the Idlib Governorate, which provoked the United States to carry out reprisal bombings against military installations of the Assad regime [18,19].

The city of Idlib remains at the nexus of converging crises and instability. Its predicament is reflective of wider trends in warfare, with a shift to urban, asymmetric forms of combat. In May 2020, the International Crisis Group

---

2. In 2020, the United Nations Security Council (UNSC) adopted resolution 2533 (2020) authorizing cross-border aid deliveries via Bab al-Hawa from Turkey into Idlib after vital aid deliveries were completely blocked [11]. Cross-border aid into Idlib started as early as January 2012, driven by local and international NGOs after the Bab al-Hawa crossing fell under opposition control. However, the UN-led cross-border response did not start until July 2014 after the UNSC adopted Resolution 2164 allowing UN agencies and their partners to deliver cross-border aid into Syria through four crossings. Since 2020, this has been whittled down to one crossing, via Bab al-Hawa from Turkey into Idlib, as per the UNSCR 2533 after Russia and China vetoed the extension of the other three crossings.

expressed concern regarding COVID-19, warning that "the global health challenge intersects with wars or political conditions that could give rise to new crises or exacerbate existing ones" [20]. Such conditions were plainly met in Idlib amid the conduct of protracted war, deteriorating violence, and humanitarian crises. The pressures on Idlib at the time of the emergence of SARS-CoV-2 were unique among global municipal centers facing the advent of a novel pandemic pathogen whilst engaged in the daily provision of care in the most strained of circumstances. Clear lessons can be drawn from this experience in relation to the vital importance of community coordination and engagement during health and other emergencies in urban conflict environments.

The public health system in Idlib

Local governance of the city of Idlib changed dramatically with the onset of the Syrian Revolution in 2011. The local revolutionary coordination committee became the foundation of many local entities that emerged to manage various aspects of daily life in the city. These entities include charities and humanitarian organizations, technical syndicates, neighborhood committees, and local administrative councils. Local Councils have played vital roles in providing municipal services, collecting data on population statistics, coordinating humanitarian interventions, and acting as an intermediary between humanitarian actors and armed groups. The first Local Council is thought to have been established in 2012 as elected, ad hoc structures to coordinate local affairs, but this approach spread across other opposition areas that came to employ similar strategies [21]. Although these Councils had limited enforcement powers, they were able to undertake core governance functions using their existing community roles and through their communication and coordination with both armed groups and humanitarian actors. There were a few leading examples in other opposition-controlled areas such as the Doma and Daryya Councils, which were reported to function better than some pre-conflict municipalities [22]. These Councils have played key roles in introducing a strong culture of democratic norms, electoral standards, and good governance to communities in opposition-held areas.

Since the city of Idlib fell under opposition control in 2015, it has witnessed multiple internal conflicts between various Islamic and opposition armed factions with repeated changes to local governance arrangements. These clashes culminated in 2019 resulting in a single dominant faction taking power. This organization, named Hayat Tahrir al-Sham (HTS), is a loose amalgamation of opposition factions, emerging from its predecessor Jabhat al-Nusra, an affiliate of al-Qaeda [23]. The organization publicly cut ties with al-Qaeda in 2017 and arrested prominent members with connections to the organization. Today HTS maintains itself to be "an independent entity that follows no organization or party, al-Qaeda or others" [24], and now casts itself as a governing opposition organization in the Idlib region, where it leads the regional military response to the ongoing conflict, as well as undertaking local government functions from economic management, law enforcement, and municipal organization. Beyond Idlib, the group has undertaken operations in Aleppo, Hama, Dera'a, and Damascus provinces.[3]

Given the complex geopolitics in the Idlib region, two main governments have claimed authority over Idlib city; the opposition Syrian Interim Government, which is based in Turkey, and the Salvation Government established by HTS. There are, however, additional local entities with control over some sectors. Local Councils, for example, have authority and wide delegation in relation to municipalities and local affairs [25]. Other technical entities without formal affiliations to either of these two governments also exist. Examples of these technical entities include the education directorate, the general directorate of water, professional syndicates, and the health directorate. Since 2019, the Salvation Government has been trying to dissolve these local technical entities and transform them into departments within their government; they have succeeded in incorporating some entities, but, at the time of writing, have not been able to absorb the city's health directorate.

A bottom-up, locally-led health system in Idlib

The main health authority in Idlib City has, since 2015, been the Idlib Health Directorate (IHD) [26]. The IHD first emerged in 2013 after several local health and medical networks came together to strengthen the coordination of the humanitarian response. This coordination was further strengthened during the polio outbreak between 2013 and 2016. This was the first major public health threat to test the readiness and responsiveness of the health system in Northwest Syria. The polio response drove the formation of both the Polio Control Task Force, as well as vaccination structures composed of wide networks of volunteers that sought to reach every household in the region. The Polio Control Task Force was the first of its kind and effectively led the response, thanks to support from various donors and enormous coordination efforts between UN agencies, the WHO, and international and local nongovernmental organizations (NGOs). In addition, considerable

---

3. Nevertheless, some entities, strongly oppose HTS' claims to power. These include entities like the USA's Department of State, which has designated HTS as a terrorist organization with residual links to al-Qaeda's leadership. The Turkish Government has also called for HTS to dissolve and join a Turkish-backed anti-Assad coalition; however, such a move has yet to occur. HTS' bid to soften its image by disavowing al-Qaeda and other Islamist group associations has also done little to deter the ongoing targeting of civilians by Assad-backed military forces.

investment was made in supporting local vaccination structures, linking these with the Health Directorates [27−29].[4]

In March 2015, Idlib City fell under opposition control, liberating the governorate of GoS forces. During the Battle of Idlib City, the IHD took a leading role in securing the health infrastructure to ensure the efficient utilization of disease control resources amid ongoing hostilities. As a result, the IHD negotiated with various armed groups involved in the conflict to obtain unified recognition of its responsibility for all hospitals, administrative buildings, equipment, and medical supplies across the governorate [30]. The IHD showed advanced health diplomatic skills through effective negotiations with various types of political and military actors convincing them of the need to merge health resources under one neutral health authority.

The central team of the IHD was designated by the relevant local authorities, including the opposition Ministry of Health (i.e., that of the Syrian Interim Government), as the key local health authority in the governorate. However, early on a need was identified to strengthen the legitimacy of the organization as the trusted provider of health services in the city. Considering the conflict sensitivities, health leaders in the Idlib Governorate attempted to orient the health sector as a neutral entity to avoid political and military interference [31].

In May 2015, the IHD called for a founding meeting, to which it invited almost all active medical doctors in the governorate to participate in the reform of the health directorate. The one-day meeting was attended by about 115 medical doctors from all subdistricts in the province and produced several key outputs [32]. First, participants agreed on the founding values of the new IHD. They then discussed ways to translate these values into practice and governance structures. It was also agreed upon that the entire health workforce; including medical doctors, nurses, midwives, and other health workers would form a general assembly for the health directorate [33]. This general assembly is represented throughout the functioning health facilities in the region. Every health facility was asked to elect a representative to take part in the election of a board of trustees. During this meeting, the participants acted as a general assembly and elected a board of eight trustees who were tasked with overseeing the work of the IHD. Last, facilitated by the eight elected trustees, the participants elected a health director for the IHD from two nominees. The health director appointed a new executive team for the IHD. The outcomes of this meeting were communicated widely to all health actors and stakeholders across the city and the wider governorate [34].

This innovative approach was developed further to accommodate feedback from the general assembly (i.e., the health workforce in the Idlib

---

4. A similar approach was later adopted for the COVID-19 response with a task force operating from Turkey, a central field coordination mechanism in Idlib, and significant accompanying investment in local volunteers' groups. This is discussed in greater detail later in this chapter.

governorate) and to accommodate for new security developments in the region. For example, an advisory board for the IHD was established in 2016 with the expansion of IHD work. The advisory board included Syrian doctors and academics from inside Syria and the diaspora who were active in humanitarian work or related public health research and practice. Moreover, the board of trustees developed a brief constitution for the IHD that defined the roles and responsibilities of each governance component in the directorate.

Following the reform of the IHD in 2015, it started to gain greater recognition from various stakeholders in the governorate. All opposition armed groups and military factions have recognized the IHD as the only health authority in the region and agreed not to interfere in their work. Humanitarian actors, including local and international NGOs, recognized the IHD as a key local health actor meriting consultation and coordination during any health project in Idlib. A group of local NGOs went even further developing a code of honor for Syrian health actors in opposition-held areas inviting all health actors to recognize the health directorates as the official technical health authorities.

In 2016 the WHO Health Cluster in Gaziantep, Turkey, also recognized the vital role the IHD played in the health sector and the IHD was invited to be an observing member of the Cluster [12]. It was also invited later to participate in the health humanitarian response plans, supplying the Cluster with situational updates from the field, and assessing proposals from NGOs on key activities [33]. The IHD invested in this opportunity and developed new plans for information sharing, and communication with all health actors, and outlined strategic proposals for health activities for the humanitarian response.

COVID-19 in Idlib

As COVID-19 spread around the world, the IHD prepared by adopting a "delay approach" to provide time for scaling-up control measures [26]. Idlib's geopolitical and territorial isolation were harnessed by local actors such as the IHD, the White Helmets, and other civil society organizations to introduce a swathe of nonpharmaceutical interventions to delay the spread of the virus into the community until public health measures could be stood up. Border closures and public awareness campaigns characterized early responses to the impending pandemic threat, whilE testing and contact tracing programs were scaled up in collaboration with the WHO-led Health Cluster based in Turkey [35]. March 3, 2020, the WHO-led health cluster established a COVID-19 Task Force consisting of local and international NGOs. The task force developed basic emergency planning for potential scenarios alongside technical guidelines [36]. However, the WHO had limited capacity to engage in such a large-scale, complex conflict environment since it did not have a physical presence inside opposition-held Syrian and had been dealing with geopolitical challenges relating to the cross-border response from Turkey via Bab al-Hawa.

The IHD, throughout its 7 years of existence, had continued to maintain its position as the technical health authority in the region. To navigate this rapidly evolving environment with a pandemic imminent, the IHD drew upon the grassroots governance systems which had been a central focus of its initial legitimacy-building and health diplomacy efforts [35].

Scaling-up health system capacity using limited resources

During the delay phase of Idlib's response, a strong collaboration developed between the IHD, the WHO health cluster task force, and Syrian diaspora medical networks in France, the United Kingdom, and the United States [26]. These strong collaborations supported the IHD in developing a clinical response plan to scale up its health services using limited resources. The plan prioritized intensive care unit (ICU) capacity, with a focus directed at scaling-up ventilator capacity where possible with the aim of responding to the predicted 14% of possible patients who would develop severe respiratory complications. At the time of SARS-CoV-2 emergence, there were only 47 adult ventilators for the whole of the Idlib region, the majority of which were located in the city itself; however, this ventilator base was expanded to around 100 by the time of the first case report some 3 months later [26,35].

Clear clinical definitions of suspected and confirmed cases were developed alongside clinical standards of procedures on how to triage, refer, diagnose, treat, and discharge patients. The task force recommended establishing three new hospitals for COVID-19 cases, but no funding was available to implement this plan. The IHD then prepared one department in the main district hospital in Idlib district to receive all suspected cases [26].

Infection prevention and control (IPC) was at the core of the early response planning. Triage tents were set up in front of most health facilities to ensure suspected patients did not infect other patients and to minimize disruption to existing routine health services [26]. The IPC guidelines and protocols were reviewed and updated to reflect the latest, rapidly evolving clinical and epidemiological evidence regarding SARS-CoV-2. In addition, more than 500 medical personnel were trained on IPC. Resource shortages nevertheless remained a critical challenge for the scale-up—despite rapid improvements in resource availability over a short timeframe, the number of ventilators, personal protective equipment, and other vital medical supplies remained insufficient for the population requirement.

The first case of COVID-19 in Northwest Syria was reported on July 10, 2020, 4 months after reporting the first case in Government controlled areas [37]. While this delay in the onset of the disease in Idlib can be attributed to many factors including the natural isolation of the opposition-controlled areas in Northwest Syria, the successful delay strategies employed by the IHD and other health actors in the region are also implicated.

There have been 2492 COVID-19-related deaths reported in the region, representing a fatality rate of 2.4%. However, this rate is also likely far higher than the actual case fatality rate considering the limited testing capacity where testing is prioritized for severe cases. The Health Information System Unit, which operates as an autonomous entity under the IHD, reports a total of 52,890 hospital admissions related to COVID-19 since the start of the pandemic up to May 31, 2022. Throughout the pandemic, the health system did not reach a maximum capacity for patient admission, including in ICU contexts. The occupation rates ranged from 13% to 97% for ICUs, and from 3% to 84% for COVID-19 wards—the highest of these occupation rates witnessed only in the last quarter of 2021 (September to December 2021), a phenomenon attributed to the Delta variant.

The COVID-19 response in Idlib

By the end of April 2020, 43 cases of COVID-19 had been reported in GoS-held areas of Syria, and none in Idlib [37]. Concern was growing that this absence of reporting was related to limited testing capacity, rather than an absence of disease, as only 1661 tests had been carried out by the end of June [26,38]. The first case of COVID-19 was officially reported in Idlib in July 2020 after a doctor tested positive [39]. At the time, a fragile ceasefire was in place in anticipation of the arrival of the disease and the ongoing demands of the existing humanitarian response. The IHD issued recommendations urging schools to close and for large gatherings including group prayers to be paused [40]. However, recommendations issued elsewhere, such as social distancing and regular handwashing, were impossible to impose in an overcrowded, impoverished, and underresourced setting after nearly a decade of war.

Community engagement and mobilizing local resources

One of the benefits of the delay approach was the ability to learn from the experiences of other settings in response to COVID-19. Evidence was growing that more effective responses were characterized by the efficient use of executive power with well-coordinated, de-centralized resource mobilization [26]. Thus the IHD, alongside other local actors including the White Helmets, the Syria Civil Defence Force, called for a mass voluntary campaign inviting local groups to take part in the response. This campaign, "Volunteers against Corona" mobilized thousands of volunteers covering most localities in Idlib [41,42]. The campaign organized volunteers in various technical teams and neighborhood committees undertaking tasks such as awareness-raising, disinfection campaigns, and community-based referrals

[26]. Neighborhood committees were responsible for raising awareness in their localities, identifying high-risk groups for shielding approaches, and linking local communities with the central campaign. Social media was employed via free cloud web servers and WhatsApp communication to reach this large group of volunteers. Facebook was used as the hub for all volunteer groups to obtain updates on the campaign and as a publishing platform for technical guidelines. WhatsApp was the platform selected for all day-to-day communication and for team management purposes.

Considering the limited capacity of the health system in Northwest Syria to deal with a possible expansion of cases, the early focus of the response was on preventative measures—including border control measures, social distancing, public awareness campaigns, disinfection campaigns, quarantine and isolation, and the use of diaspora networks—to contain and delay the spread of the virus (Table 4.1) [26,30,33].

Control of border crossings and other points of entry proved an essential component of the delay phase of Idlib's response, although in the longer-term key challenges have emerged in relation to the longevity of Bab al-

TABLE 4.1 Key COVID-19 preventative measures in Northwest Syria [26,34,35].

Measure	Approach for implementation
Border control measures	All official crossing points with the Government of Syria and Northeast Syria were closed from mid-March 2020. Only essential medical evacuation and humanitarian aid deliveries were permitted via Bab al-Hawa.
Social distancing	All health actors asked people to stay home, where possible, and to reduce social gatherings and events. All schools were closed.
Public awareness campaigns	Health and local nongovernmental organizations engaged in various public awareness activities including distribution of more than a million educational materials (e.g., leaflets and brochures), household visits in camps, and radio messaging.
Disinfection campaigns	These campaigns targeted mainly the residential collective centers, camps, and public buildings (e.g., schools and health facilities).
Quarantine and isolation	The Idlib Health Directorate established 17 Community-Based Isolation centers with a capacity of 1400 beds for opening May 2020.
Use of diaspora networks	Syrian health and medical diaspora networks from France, the United Kingdom, and the United States provided training, evidence updates, and clinical input via digital platforms (e.g., Violet Syria).

Hawa as a viable point of entry for humanitarian supplies. On March 15, 2020 all crossings between the Idlib region and GoS regions were closed in addition to others in opposition-controlled Northeast Syria [31]. The Bab al-Hawa border crossing with Turkey was restricted from the Turkish side with cross-border activities only permitted for essential medical, commercial, and humanitarian aid deliveries. The UN Security Council passed Resolution 2504 in January 2020, extending cross-border operations via this sole remaining point of entry to avoid the total collapse of necessary supplies during the pandemic [43]. However, border control measures seen in other countries, such as quarantine of delivery drivers, could not be implemented in the absence of enforcement structures and given the critical dependence on incoming supplies.

Stay-at-home guidelines were issued by medical professionals and the IHD; however, nonpharmaceutical interventions such as lockdowns and social distancing measures were simply not feasible in Idlib given the precarity of local populations reliant on unstable work and humanitarian aid to meet the most basic of subsistence needs. Populations in camps living in tents could not be confined to their homes given overcrowded conditions. The need for communities to line up for water and food supplies, as well as limited access to sanitation and hygiene facilities made the standard nonpharmaceutical interventions deployed around the world impossible to implement. These considerations are unique to the context of war and humanitarian crisis and offer painful lessons for emergency planners such as the IHD tasked with multiple, overlapping crises. Further, other important dimensions soon became apparent including the important role of culture and ritual in the daily conduct of communities, as witnessed in prior health emergencies, such as Ebola [44]. In particular, funereal and prayer practices involving large gatherings were a challenge for authorities to address [1,40]. Close engagement between Local Councils and religious leaders formed the bedrock of risk mitigation around these practices. During high peaks of COVID-19, religious gatherings, such as Friday prayers, were partially suspended or reduced in numbers. Similarly, funeral practices were amended to restrict contact with corpses and limitations on attendee numbers were imposed.

Multiple international groups collaborated with the IHD on COVID-19-focused public awareness campaigns. Prior to the pandemic, a European Union-funded initiative named the Community Cohesion and Stability Project established a training program involving informal local community groups called *Tawafoq* Initiatives [45]. These groups were set up to engage on issues pertinent to decision-making for vulnerable communities, particularly IDPs and women. However, in response to the unfolding COVID-19 pandemic, these *Tawafoq* Initiatives joined forces and set up a joint program called "Together Against Coronavirus." Working with the IHD, the White Helmets, and Local Councils, the public awareness campaign that emerged from this effort delivered in-person awareness-raising activities on issues such as self-isolation, case recognition, hygiene, and disinfection [41]. Further funding supported the production of videos, painting of educational murals, and the placement of posters across the city. Local media outlets publicized this work heavily and 95% of residents reported having seen the initiative's outputs in public [45]. These and other volunteer-driven public awareness campaigns formed the backbone of public health education over the course of the pandemic. Given the constrained health workforce, civilian involvement or task shifting in health education was a critical local dimension of response capacities.

The White Helmets, the Syrian Civil Defense, force played a key role in public engagement around the coronavirus pandemic in Idlib. Also, closely involved in public awareness campaigns, the organization provided daily updates via Facebook and other social media channels [41]. Consisting of around 3000 volunteers, it also launched a public disinfection and prevention campaign. Over 510 sites were disinfected 2020, including 405 medical facilities, 1515 mosques, 1159 schools, and 693 IDP camps [46]. The group ran over 300 health education sessions on the virus, while maintaining its prior functions including clearance of unexploded ordnance and rebuilding municipal infrastructure such as roads.

The IHD established 17 community isolation centers, including for cross-border travelers returning home from Turkey via Bab al-Hawa [47]. The objective was to establish 1400 additional beds for community-based isolation to alleviate pressure on the existing health and humanitarian infrastructure [48]. Additional input from volunteer groups to support referral, transfer, and daily support of isolation procedures was essential for maintaining the functions of these newly constructed sites. The number of these community isolation centers increased in 2021 as they played an effective role in reducing the burden on the local health system. Most mild and moderate cases were managed within these centers without the need to visit formal health facilities, an innovation that also helped to reduce disease transmission. In 2022 the need for these centers decreased drastically and according to the Health Information System Unit—the occupation rate of these centers has dropped from 64% in October 2021 to 17% in January 2022, to less than 1% since April 2022.

With the novelty of SARS-CoV-2 driving an explosion of global scientific research, the rapidly changing evidence base proved a challenge across nations and cities worldwide when it came to tailoring response interventions to public health predictions. This dimension of the response in Idlib drew upon the wide geographic dispersal of the Syrian health and medical community around the world. Communities in France, the United Kingdom, and the United States worked with the IHD and local volunteer groups to disseminate up-to-date and emerging evidence and to advise on intervention protocols [25]. A central medical chat room was established on WhatsApp where relevant information was shared on a daily basis. Several online remote training sessions were provided using a variety of online platforms including Skype, Zoom, GoTo Meeting, and Google Meet. In addition, a repository of resources, educational materials, and training packages was collated and disseminated across these diverse channels for use by field health workers.

SARS-CoV-2 testing and COVID-19 surveillance

Diseases surveillance in all territories outside the control of the GoS is conducted by the Early Warning and Response Network (EWARN)—a syndromic surveillance system for communicable diseases established by the Assistance Coordination Unit, which is a Syrian NGO that was established in 2013 [26,30,48]. Shortly after its establishment, the EWARN was able to share the initial alert for the first case of polio in eastern Syria in October 2013.

At the outset of the pandemic, polymerase chain reaction (PCR) testing capability was very limited in Northwest Syria. Since 2017 EWARN had been working on scaling-up its capacity to identify two priority pathogens: severe acute respiratory infection and influenza-like illness [29]. With technical and financial support from the Gates Foundation, EWARN had placed a request with the WHO to supply it with the required training and equipment to develop a reference lab for PCR testing inside Syria. As a result, with the start of the flu season in September 2019, EWARN had started preparing for a "flu-like" disease outbreak. This preparation entailed refining the reporting procedures—including revisiting definitions and communication channels, providing training refreshers for health and medical practitioners and field officers, and communicating with regional and international health actors for technical and resource support. This preparation meant that the EWARN maintained the basic capacity to deal with the first phase of the COVID-19 response. As early as November and December 2019, the EWARN had identified an outbreak of H1N1 in northern Aleppo, and procedures were revised during the "delay" phase of Idlib's COVID-19 experience to support verification, triaging, and testing procedures for suspected cases of COVID-19 [47].

At the start of the COVID-19 pandemic, EWARN quickly ran out of its limited supply of 300 PCR tests and relied on a Turkish reference lab to supply the required materials and training to conduct a very limited number of tests. By mid-April, EWARN had received another shipment of PCR tests from the WHO, with an additional 5000 tests provided [47]. Still, as of April 22, 2020, only 191 COVID-19 tests had been carried out across Northwest Syria, and only 20 tests per day were being processed in the single laboratory capable of conducting PCR testing [49]. During the delay phase and beyond, EWARN worked to scale up this capacity to a target of 100 tests per day and had developed the capacity to conduct a few thousand PCR tests by August 2020.

The EWARN has also been proactive in compensating for the absence of robust surveillance systems by assuming the functions necessary for the surveillance and modeling of COVID-19. It has developed a reporting system with robust field communication channels established with all facilities that provide COVID-19-related services. And, since June 2020, the EWARN has been producing daily reports on COVID-19 cases, tests, and deaths. These reports are shared with the WHO Health Cluster in Gaziantep and with the wider health actors in the region. This information was instrumental for health actors to monitor the pandemic and plan the ongoing response.

As of July 7, 2022, according to the EWARN, there have been 103,004 cases of COVID-19 in Northwest Syria reported through 394,766 PCR tests, representing a test positivity rate of 26%.[5] This high test positivity rate demonstrates the limited testing capacity and indicates that the actual number of cases is likely to be far greater than the reported numbers. This is supported by the increased incidence of influenza-like illness reported through the EWARN in 2020 and 2021, which might indicate increases in COVID-19 cases that were not reported due to limited testing capacity.

In addition to the testing services provided throughout the pandemic, the EWARN has been responsible for a wide range of public health activities including contact tracing and risk mapping, as well as the introduction of measures related to hygiene, health promotion, water and sanitation [26,47].

Resource-limited digital solutions

The expansion of digital technologies in health care (eHealth) has been observed throughout the global COVID-19 experience [50]. However, even prior to the emergence of the virus, the rollout of these technologies in conflict settings had been identified as a potential tool for easing the constraints of limited resources, health worker insecurity, and geographic obstacles. Mobile applications had been shown to provide valuable early syndromic data for health care planners in settings such as the United Kingdom. The development of the ZOE app at King's College London provided the first evidence-supported finding that anosmia was a key predictive symptom of SARS-CoV-2 infection in April 2020 [51]. Building on these insights, Syrian medical experts from the United Kingdom diaspora worked on the development of an Arabic language online self-assessment tool for delivery in Idlib and wider Northwest Syria.

The website collected information on several key parameters; (1) users' location localized to subdistrict level, (2) travel history, (3) suspected signs and symptoms, (4) risk factors, and (5) the presence of severe complications.

---

5. These estimates are sourced from internal reporting datasets of EWARN.

Based on this information, an algorithm classified users into five categories depending on the level of disease suspicion and the presence of any risk factor or complication that might require immediate attention. Users were advised according to their responses on appropriate self-referral and health care behaviors. The website was developed in coordination with a local Syrian NGO called "Violet Syria" which had been involved in various social mobilization activities related to COVID-19 [26]. The organization has coordinated with local groups to advise communities to use the website for initial self-assessment with the aim of reducing pressure on an already stretched health system. Dissemination of the tool by local volunteer groups working on the ground proved critical to supporting the effective rollout of the tool. According to Violet Syria, this tool provided more than 2 million checks in the first year of the pandemic to communities mainly in Idlib and Northwest Syria but also in Turkey and other neighboring countries. However, with the increased availability of other local services related to COVID-19, and with the rapid changes in the clinical presentation of cases, the relevance of this diagnostic tool became limited in 2021 and was subsequently suspended in late 2021.

Social media and communication tools have been recruited effectively in collecting health information and disseminating important health messages. The IHD has dedicated a WhatsApp number to respond to queries from the public on COVID-19. Syrian diaspora networks have also been involved in providing remote consultations and support to members of the public as well as to their medical colleagues on the ground. Considering the widespread use of Facebook in Northwest Syria, Facebook has been effective in organizing volunteer campaigns and for conveying messages. In the absence of central health information infrastructure, these tools have proved vital to the daily running of the local response to the COVID-19 health emergency. The wider global experience of the pandemic has made it clear that adaptable responses to COVID-19 have required an adaptable accompanying information architecture. Despite impressive local responses to this challenge during COVID-19 and prior emergencies, such as polio, a longer-term requirement to strengthen health information systems is critical to delivering resilient health services in the context of overlapping crises.

Vaccination

COVID-19 vaccination in Idlib Governorate has been delivered by COVAX, the vaccine pillar of the Access to COVID-19 Tools Accelerator—an effort led by the WHO, the Coalition for Epidemic Preparedness Initiative, and GAVI [52]. The first shipment of 53,800 doses arrived in the city in April 2021 via the Bab al-Hawa crossing and initial vaccination efforts prioritized health care workers [53]. These efforts were overseen by the Syria Immunization Group, which worked closely with the WHO Cluster in

Gaziantep to implement COVID-19 vaccination campaigns within the governorate. Numerous NGOs pooled funding to support the Syria Immunization Group plans. Still, similar to other low-income countries, there have been limited and delayed supplies available to support the vaccination campaign in Idlib, which have been further compounded by uncertainty regarding the flow of aid through the Bab al-Hawa crossing.

Conclusion

The conditions in Idlib remain precarious given the ongoing humanitarian crisis, widespread poverty, and enduring violence. At the time of writing, the Russian Government has exercised its veto at the UN Security Council and the crossing at Bab al-Hawa is yet again subject to restrictions on cross-border aid transfers. Of all the crises to befall Idlib in the last 11 years, COVID-19 is but one episode in a wider story of recurrent tragedy.

Several critical needs have been identified from Idlib's COVID-19 experience. There is a need to expand surveillance tools, including scaling-up laboratory capacities and exploring the role of event-based systems, in addition to other, more traditional approaches. Local ownership and priority setting have shown their worth. However, without further support from the international community in this enormously complex political setting, gains are unlikely to be sufficient or sustainable in the face of ongoing conflict. The relative absence of support from international actors despite the continued hostility from the GoS remains a handbrake on regional and municipal recovery in the health sector and beyond.

Among health practitioners in Idlib, there is a sense that COVID-19 spared the region of its worst ravages. Much uncertainty remains as to why the health sector was not overwhelmed, despite the worst predictions. Clearly, lessons had been learned from prior health emergencies, such as polio, about the power and importance of local-level coordination and community engagement via civil society groups, volunteer organizations, and local health authorities. The establishment of quarantine and isolation centers was a valuable buffer to support an already overburdened health sector. There is, likely, a behavioral element also, which may explain why hospital admissions did not reach unsustainable levels; how health-seeking behaviors are altered in conditions of conflict and overlapping crises remains an important question for both academia and health care providers to consider and a wider program of research can inform this line of inquiry.

The value of local health leadership has been critical for the city of Idlib during its COVID-19 response. However, the health sector remains in crisis given the catastrophic degradation of human capital and material resources. Diaspora medical networks have been extremely agile throughout the conflict—using knowledge networks and eHealth tools to great effect. The legacy of the Arab Spring has shown itself in the tremendous social consequence of social media tools such as WhatsApp across this region. There do, however, remain serious challenges for the ongoing provision of medical care by trained providers—the loss of health workers places Idlib at great risk of future health emergencies, whether they are future waves of COVID-19 or the growing threat of antimicrobial resistance. Supporting conflict cities to face these sequential threats should be a key imperative for health planners as war and insecurity continue to coincide with the perpetual threat of infectious disease.

References

[1] Alkarim T, Megally H, Zamore L. Last Refuge or Last Hour? COVID-19 and the Humanitarian Crisis in Idlib. NYU Centre for International Collaboration. 2020 https://cic. nyu.edu/sites/default/files/idlib-covid-19-briefing-web-final.pdf - last accessed 7/7/22.

[2] France 24. Idlib: Syria's 'forgotten city' targeted for revenge by Assad. 13th March 2020. https://www.france24.com/en/20200313-idlib-syria-s-forgotten-city-targeted-for-revenge-by-assad - last accessed 7/7/22.

[3] France 24. Civilians killed as fresh violence erupts in Syria. 3rd December 2011. https:// www.france24.com/en/20111203-deadly-violence-syria-authorities-battle-army-defectors-united-nations-civil-war-bashar-al-assad - last accessed 7/7/22.

[4] Laub Z. Syria's Civil War: The Descent into Horror. Council on Foreign Relations 17th March 2021. https://www.cfr.org/article/syrias-civil-war - last accessed 7/7/22.

[5] ITV. Crisis in Syria reaches 'horrifying new level' as UN warns of 'biggest humanitarian horror story of 21st Century'. 17th February 2020. https://www.itv.com/news/2020-02-17/ crisis-in-syria-reaches-horrifying-new-level-as-un-warns-of-biggest-humanitarian-horror-story-of-21st-century - last accessed 7/7/22.

[6] UNHCR. UN High Commissioner for Refugees appeals for safety for civilians trapped in Idlib. 20th February 2020. https://www.unhcr.org/uk/news/press/2020/2/5e4e51d04/un-high-commissioner-refugees-appeals-safety-civilians-trapped-idlib.html - last accessed 7/7/ 22.

[7] Amnesty International. Syria: Non-renewal of last aid corridor risks humanitarian catastrophe for millions. July 5th 2022. https://www.amnesty.org/en/latest/news/2022/07/syria-non-renewal-of-last-aid-corridor-risks-humanitarian-catastrophe-for-millions/ - last accessed 7/7/22.

[8] Duclos A, Ekzayez A, Ghaddar F, Checchi F, Blanchet K. Localisation and Cross-Border Assistance to Deliver Humanitarian Health Services in North-West Syria: A Qualitative Inquiry for *The Lancet*-AUB Commission on Syria'. Conflict and Health 2019;13:20.

[9] United Nations Security Council. Security Council resolution 2533 (2020) [on humanitarian situation in the Syrian Arab Republic and renewal of authorization of relief delivery and monitoring mechanism for a period of 12 months]; 2020. https://digitallibrary.un.org/ record/3871951?ln=en — last accessed 7/7/22.

[10] Ekzayez A. Attacks on healthcare in the Syrian conflict: Drawing lessons from Syria to improve global reporting systems. London: Chatham House; 2021. Nov 23.

[11] ICRC. Health-care providers, patients suffer thousands of attacks on health-care services over the past five years. 3rd May 2021. https://www.icrc.org/en/document/health-care-providers-patients-suffer-thousands-attacks-health-care-services-past-5-years - last accessed 7/7/22.

[12] Fouad FM, Sparrow A, Tarakji A, et al. Health workers and the weaponisation of health care in Syria: a preliminary inquiry for the Lancet-American University of Beirut Commission on Syria. Lancet 2017;390:251626.

[13] Philp C. Idlib will be a bloodbath, warns UN. 25h February 2020. https://www.thetimes. co.uk/article/idlib-will-be-a-bloodbath-warns-un-zjn7rh8bv - last accessed 7/7/22.

[14] UOSSM Press Release. 10 Schools and Hospital Bombed in Idlib Today, Killing Children and Teachers. 25th February 2020. https://reliefweb.int/report/syrian-arab-republic/10-schools-and-hospital-bombed-idlib-today-killing-children-teachers - last accessed 7/7/22.

[15] Gharibah M, Mehchy Z. COVID-19 Pandemic: Syria's Response and Healthcare Capacity. LSE Conflict Research Programme Memo; 25th March 2020. https://www.lse. ac.uk/school-of-public-policy/assets/Documents/Social-Sciences-Response-to-Covid/Dr-Rim-Turkmani.pdf - last accessed 7/7/22.

[16] Ensor J. Britain and US threaten response after reports of fresh chemical weapons attacks in Syria. The Times, 22nd May 2019. https://www.telegraph.co.uk/news/2019/05/22/brit-ain-us-threaten-response-reports-fresh-chemical-weapons/ - last accessed 7/7/22.

[17] Organisation for the Prohibition of Chemical Weapons. Note by the technical secretariat - Report of the OPCW Fact-Finding Mission in Syria Regarding Alleged Incidents in the Idlib Governorate of the Syrian Arab Republic between 16 March and 20 May 2015. S/1319/2015; 29th October 2015.

[18] Organisation for the Prohibition of Chemical Weapons. Note by the technical secretariat − Report of the OPCW Fact-Finding Mission in Syria Regarding an Alleged incident in Khan Skaykhun, Syrian Arab Republic. S/1510/2017; 29th June 2017.

[19] BBC. Syria air strikes: Trump hails 'perfect' mission. 14th April 2018. https://www.bbc. co.uk/news/world-middle-east-43767156 - last accessed 7/7/22.

[20] International Crisis Group. COVID-19 and conflict: seven trends to watch [Internet]. *Crisis Group Special Briefing N°4*. 2020. https://www.crisisgroup.org/global/sb4-covid-19-and-conflict-seven-trends-watch - last accessed 7/7/22.

[21] Hajjar B, von Burg, Hilal L, Santschi M, Gharibah M, Sharbaji M. Perceptions of gover-nance- The experience of local administrative councils in opposition-held Syria. Bern: SwissPeace; 2017.

[22] Angelova I. Governance in rebel-held East Ghouta in the Damascus Province, Syria. CGHR Working Paper 10. Cambridge: University of Cambridge Centre of Governance and Human Rights; 2014.

[23] Zelin A. The Age of Political Jihadism: A Study of Hayat Tahrir al-Sham. Policy Focus 175, The Washington Institute for Near East Policy; 2022.

[24] Center for Strategic and International Studies. Hay'at Tahrir al Sham (HTS). 4th October 2018. https://www.csis.org/programs/transnational-threats-project/past-projects/terrorism-backgrounders/hayat-tahrir-al-sham - last accessed 7/7/22.

[25] Al-Achi A. Civil Society is the Last Line of Defence Against HTS in Northwest Syria. Chatham House; 2020. https://syria.chathamhouse.org/research/civil-society-is-the-last-line-of-defence-against-hts-in-northwest-syria - last accessed 7/7/22.

[26] Ekzayez A, Al-Khalil M, Jasiem M, Al Saleh R, Alzoubi Z, Meagher K, Patel P. COVID-19 response in northwest Syria: innovation and community engagement in a complex con-flict. J Public Health (Oxf) 2020;42(3):504−9.

[27] Al-Moujahed A, Alahdab F, Abolaban H, Beletsky L. Polio in Syria: Problem still not solved. Avicenna J Med 2017;7:64−6.

[28] Ismail SA, Abbara A, Collin SM, et al. Communicable disease surveillance and control in the context of conflict and mass displacement in Syria. Int J Infect Dis 2016;47:15−22.

[29]   Ahmad B, Bhattacharya S. Polio eradication in Syria. Lancet Infect Dis 2014;14:547—8.

[30]   Abbara A, Ekzayez A. Healthcare leadership in Syria during armed conflict and the pandemic. BMJ Global Health 2021;6:e005697.

[31]   Beaujouan J. The COVID-19 Pandemic and Alternative Governance Systems in Idlib. IDS Bulletin 2022;53(2).

[32]   Ekzayez A. Analysis: A Model for Rebuilding Infrastructure in Northwestern Syria. The New Humanitarian. 2018. https://deeply.thenewhumanitarian.org/peacebuilding/articles/2018/02/19/analysis-a-model-for-rebuilding-infrastructure-in-northwestern-syria - last accessed 23/10/22.

[33]   Alzoubi Z, Iyad K, Othman M, Alnahhas H, Abdulaziz Hallaj O. Reinventing State: Health Governance in Syrian Opposition-Held Areas. MIDMAR. 2019.

[34]   Ekzayez A. (2023) *Health system strengthening in conflict settings: a case study of a hybrid health system in northwest Syria between 2013-2021* [Unpublished Doctoral Dissertation]. King's College London.

[35]   Abbara A, Rayes D, Fahham O, Alhiraki OA, Khalil M, Alomar A, Tarakji A. Coronavirus 2019 and health systems affected by protracted conflict: The case of Syria. Int J Infect Dis 2020;96:192—5.

[36]   World Health Organization (WHO), Office for the Coordination of Humanitarian Affairs (OCHA). SYRIAN ARAB REPUBLIC Whole of Syria COVID-19 Response Update No.01. 2020. https://reliefweb.int/sites/reliefweb.int/files/resources/COVID-19_Update_Issue01.pdf — last accessed 7/7/22.

[37]   UNOCHA Syria. Syrian Arab Repulic: COVID-19 Update No. 7 — 25 April 2020. https://reliefweb.int/report/syrian-arab-republic/syrian-arab-republic-covid-19-update-no-07-25-april-2020 - last accessed 23/10/22.

[38]   Save the Children. First COVID-19 Case in North West Syria: Fears of Raid Outbreak in this Densely Populated Area. 9th July 2020. https://www.savethechildren.net/news/first-covid-19-case-north-west-syria-fears-rapid-outbreak-densely-populated-area - last accessed 7/7/2022.

[39]   BBC. Coronavirus: Idlib's first COVID-19 case raises fears for Syria camps. 10th July 2020. https://www.bbc.co.uk/news/world-middle-east-53358569 - last accessed 7/7/22.

[40]   Al-Modon. *Idlib: Suspending Friday Prayers… Threatens a Rift Between Salafis*, 3 April 2020 (in Arabic). https://www.almodon.com/arabworld/2020/4/3/تعليق-صلاة-الجمعة-بإدلب-هي-ددخرش-بين-نيين-السلافيين - last accessed 7/7/22.

[41]   Enab Baladi. 'Volunteered against Coronavirus'… Solidarity in northern Syria to fight Coronavirus crisis. 2020.

[42]   Najjar F. Rescuers in war-torn Syria gear up for coronavirus battle. Al Jazeera. 2020. https://www.aljazeera.com/news/2020/03/rescuers-war-torn-syria-gear-coronavirus-battle-200325093458400.html - last accessed 7/7/22.

[43]   WHO. *Health Cluster Bulletin: January 2020*, Gaziantep: Health Cluster Turkey Hub, World Health Organization 2020.

[44]   Abramowitz SA, McLean KE, McKune SL, et al. Community-Centered Responses to Ebola in Urban Liberia: The View from Below. PLoS Negl Trop Dis 2015;9.

[45]   Adam Smith International. Supporting Community Responses to COIVD-19 in North-West of Syria. 2021. https://adamsmithinternational.com/app/uploads/2021/01/Case-Study-Tawafoq-Initiatives.pdf - last accessed 7/7/22.

[46]   UK Parliament. The impact of coronavirus on Syria. A report by The Syria Campaign for the International Development Committee's inquiry 'Humanitarian crises monitoring: impact of coronavirus'. 16th April 2020. https://committees.parliament.uk/writtenevidence/1868/html/ - last accessed 7/7/22.

[47] Syria Public Health Network. Policy Report: COVID-19 situation in Syria and possible policy responses. https://r4hc-mena.org/wp1/wp-content/uploads/2020/05/PolicyBrief-COVID19-Syria_29.4.2020_FINAL-1.pdf - last accessed 7/7/22.

[48] Diggle E, Welsch W, Sullivan R, et al. The role of public health information in assistance to populations living in opposition and contested areas of Syria, 2012-2014. Confl Health 2017;11:33.

[49] NW Syria Task Force. Coronavirus disease 2019 (COVID-19) Situation Update As of 22 April, 2020. HumanitarianResponse. 2020. https://www.humanitarianresponse.info/en/operations/stima/document/coronavirus-disease-2019-covid-19-nw-syria-task-force-situation-update-22 - last accessed 7/7/22.

[50] Bowsher G, El Achi N, Augustin K, Meagher K, Ekzayez A, Roberts B, Patel P. eHealth for service delivery in conflict: a narrative review of the application of eHealth technologies in contemporary conflict settings. Health Policy and Planning 2021;36(6):974−81.

[51] King's College London. ZOE COVID Study app: How King's researchers slowed the spread of COVID-19. 7th May 2022. https://www.kcl.ac.uk/news/spotlight/zoe-covid-study-app-kings-researchers-slowed-the-spread-covid-19 - last accessed 7/7/22.

[52] Syria's Idlib Region to Receive first batch of vaccines. Al Jazeera, 21 April 2021. https://www.aljazeera.com/news/2021/4/21/syrias-idlib-region-to-receive-first-batch-of-covid-19-vaccines - last accessed 24/10/22.

[53] World Health Organisation. Update on COVID-19 Vaccination in Syria, June 14, 2021. https://www.emro.who.int/syria/news/update-on-covid-19-vaccination-in-syria.html - last accessed 24/10/22.

Section II

Technology and digital approaches for pandemic response

What are digital approaches?

Technology played a vital role in the pandemic response by facilitating human connection amid physical distancing requirements and supporting the clinical and public health responses. Virtual conferencing, banking, and other technologies reduced the necessity for in-person work. Public health departments disseminated risk communications materials via social media and conducted contact tracing via mobile phone apps, health care practices scaled up telemedicine services, and governments distributed digital vaccine certificates [1−3]. These technologies ensured access to basic services during lockdowns and supported COVID-19 mitigation efforts.

How have digital approaches been implemented well?

Citizen trust in government and data privacy standards plays a key role in community uptake of digital health technologies. Many governments struggled to achieve widespread use of contact tracing and quarantine monitoring apps due to data privacy concerns, which were sometimes well-founded [4,5]. Singapore experienced a high degree of uptake, but as described in the following section, had to implement data privacy legislation to achieve compliance. Many countries eventually turned to the decentralized Exposure Notification System API developed by Google and Apple [6−8]. Faith in data privacy was instrumental because contact tracing apps and vaccination required widespread uptake to effectively mitigate disease spread.

System interoperability was also vital for digital health technologies [8]. Contact tracing apps were less effective if they did not track people across subnational or even national borders. Ensuring electronic health and laboratory records were easily transmitted across care facilities was crucial for effective patient care—a challenge noted by health authorities in Quezon in the following chapter [9].

Why do digital approaches matter?

Digital solutions significantly reduced demands on public health departments and hospital administrators. For instance, effective, manual contact tracing required a large workforce, which were overwhelmed during case spikes [2,10]. Many countries had to mobilize students, the military, or other workforces to support contact tracing efforts [2,11]. One notable example during the response to the COVID-19 pandemic was seen in France where the national health insurance fund reassigned full-time employees to contact tracing in the fall of 2020, helping to bring their total staff to more than 12,000 contact tracers [11]. Automated contact tracing, on the other hand, lifted some of this burden and allowed public health departments to prioritize other aspects of the response or routine public health functions. Furthermore, when case totals exceeded contact tracing staff capacity, contact tracing apps allowed communities to protect themselves. However, these digital solutions had significant limitations; poor uptake reduced efficacy, and the invasive nature of location tracking led to privacy concerns, particularly when apps were paired with facial recognition and GPS technologies [12].

Many governments and institutions also used technology and digital approaches to aid in the allocation of vaccines—using algorithms and large-scale analysis of patient medical records [13,14]. By automating these functions, public health officials and hospital administrators were able to better focus on community engagement, policymaking, patient care, and other key outbreak response activities.

How do digital approaches help in urban settings?

Managing clinical resource distribution, patient care, and contact tracing was especially challenging in areas with high population densities. Contact tracing staff in Contact tracing staff in large cities, such as Singapore, became quickly overwhelmed as cases spiked [15]. Automating contact tracing and resource allocation reduced the burden on public health staff so that they more efficiently addressed competing priorities [15]. Expanding telemedicine and automating certain hospital processes helped reduce the burden on strained health care providers, allowing them to see patients more quickly and focus on patient care rather than resource allocation.

Chapter introductions

The chapters in this section of the book detail how two large urban environments—Singapore and Quezon City, Philippines—integrated novel technologies and digital approaches into their COVID-19 responses.

Singapore, a highly urbanized island state, used a variety of technologies to mount its pandemic response (Fig. 1). Partnerships between the government, the public and private sectors, and the population have enabled authorities to rapidly mobilize public health services to respond to the emergence of SARS-CoV-2. The response mounted in Singapore is one that has drawn heavily on core public health functions and capacities such as surveillance and testing, contact tracing, isolation and quarantine, and vaccination, which were increased and adapted to meet the demands of a rapidly evolving virus and crisis—allowing the country to achieve one of the highest vaccination rates and lowest mortality rates globally.

Similarly, Quezon City, Philippines used technology to bolster its pandemic response (Fig. 2). Building upon a foundation provided by strong leadership and aspirations to use the emergency to prepare the city for the future, authorities promoted a digitization of government records and a new digital citizen identification system "QCitizen ID," designed to streamline the distribution of licenses and government aid, respectively. While these innovations involved substantial human and capital investments, they also will leave the city better prepared to respond to future public health emergencies, like epidemics and pandemics.

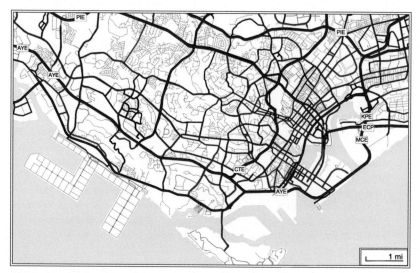

FIGURE 1 A map of Singapore. Base map sourced from OpenStreetMap, using materials made available under CC BY-SA 2.0.

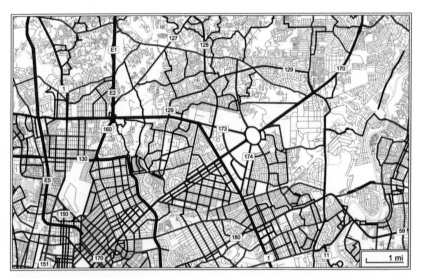

FIGURE 2 A map of Quezon City, Philippines. Base map sourced from OpenStreetMap, using materials made available under CC BY-SA 2.0.

References

[1] Clipper B. The influence of the COVID-19 pandemic on technology: adoption in health care. Nurse Leader 2020;18(5):500−3. Available from: https://doi.org/10.1016/j.mnl.2020.06.008.

[2] Moore M, Toole K. Contact tracing during the COVID-19 pandemic: a systematic comparison of global practices and influencing factors. Washington, DC: Georgetown University Center for Global Health Science & Security; 2021.

[3] Wosik J, Fudim M, Cameron B, Gellad ZF, Cho A, Phinney D, et al. Telehealth transformation: COVID-19 and the rise of virtual care. Journal of the American Medical Informatics Association 2020;27(6):957−62. Available from: https://doi.org/10.1093/jamia/ocaa067.

[4] Fowler GA. One of the first contact-tracing apps violates its own privacy policy. Washington Post 2020;21 May.

[5] News BBC. Coronavirus: Israel halts police phone tracking over privacy concerns. BBC News 2020;23 Apr.

[6] Sabbagh D, Hern A. UK abandons contact-tracing app for Apple and Google model. The Guardian 2020;18 Jun.

[7] Rahman M. Here are the countries using Google and Apple's COVID-19 Contact Tracing API. *XDA Developers*; 2021

[8] Greenberg A. State-based contact tracing apps could be a mess. Wired 2020;27 May.

[9] Greene DN, McClintock DS, Durant TJS. Interoperability: COVID-19 as an Impetus for Change. Clinical Chemistry 2021;67(4):592−5. Available from: https://doi.org/10.1093/clinchem/hvab006.

[10] Stein R. Pandemic is overwhelming U.S. public health capacity in many states. What Now?" NPR 2020;28 Jul.

[11] Williams GA, Maier CB, Scarpetti G, Galodé J, Lenormand MC, Ptak-Bufken K, et al. Human resources for health during COVID-19: creating surge capacity and rethinking skill mix. Eurohealth 2022;28(1):19−23.

[12] Whitelaw S, Mamas MA, Topol E, Van Spall HGC. Applications of digital technology in COVID-19 pandemic planning and response. Lancet Digital Health 2020;2(8):e435−40. Available from: https://doi.org/10.1016/S2589-7500(20)30142-4.

[13] COVAX WHO. Allocation logic and algorithm to support allocation of vaccines secured through the COVAX Facility. Geneva: COVAX;; 2021.

[14] Singer N. Where do vaccine doses go, and who gets them? The algorithms decide. The New York Times 2021;07 Feb.

[15] Mueller UE, Omosehin O, Akinkunmi AE, Ayanbadejo JO, Somefun EO, Momah-Haruna AP. Contact tracing in an African megacity during COVID 19: Lessons learned. African Journal of Reproductive Health 2020;24(2):27i31. Available from: https://doi.org/10.29063/ajrh2020/v24i2s.4.

Chapter 5

Singapore's whole-of-nation strategy for pandemic response and vaccination of the population

Wycliffe Wei[1], Bryan W.K. Chow[1], Marc Ho[1,2] and Vernon Lee[1,2]

[1]*Ministry of Health, Singapore, Singapore,* [2]*Saw Swee Hock School of Public Health, National University of Singapore, Singapore, Singapore*

Background

Singapore has responded to the COVID-19 pandemic comprehensively, initially pursuing a stringent containment strategy that saw the intensive mobilization of resources to test, trace, and isolate cases, and quarantine contacts. For 2020 and most of 2021, the country adopted this containment strategy, similar to many other countries during the initial phase of the pandemic. The availability of vaccines ushered in another phase of response, where central provisioning and distribution of COVID-19 vaccines supported vaccinating the entire population. However, novel variants of concern, particularly Delta and Omicron, thwarted hopes of an easy transition even after achieving high vaccination rates. Given this, Singapore cautiously persisted with adapted disease containment measures until the end of 2021, as cases of the Delta variant accrued and so did understanding of its profile. Subsequently, some disease mitigation measures continued to be enforced until the end of the Omicron BA.1/2 wave toward the end of April 2022 and were progressively relaxed through the subsequent Omicron BA.4/5 and XBB waves. Pandemic-related measures were fully lifted in February 2023.

Singapore has seen low rates of COVID-19 morbidity and mortality when compared internationally. As of June 2022, there were a total of 1300 COVID-19-related deaths in a population of around 5.5 million (or, around 240 reported COVID-19 deaths per million population). Singapore's initial stringent containment strategy contributed to these relatively low rates, which gradually transited toward a disease mitigation approach after high vaccination coverage was reached in late 2021. By the end of 2021, Singapore had one of the highest vaccination rates and the lowest excess mortality rates in the world.

The key to any strategy is the ability to realize its aims through the actual implementation of measures and the provision of services. The fast-moving local and international situation necessitated an adaptable response with an ability to nimbly affect measures and services. Singapore was able to achieve these through governmental action and through the successful engagement of the private sector and the population.

This chapter describes and discusses how Singapore responded to the pandemic through a whole-of-society approach. It first describes the context and background of Singapore's public health and crisis response system, before going through the major domains of COVID-19 response, namely: surveillance and testing, contact tracing, isolation and quarantine, and vaccination. The key lessons from Singapore's response are then discussed, including how they may contribute toward pandemic preparedness in the future.

Background on Singapore

Singapore is an island city-state in Southeast Asia that is highly urbanized, with a population of 5.6 million residents. The high human density and mobility in Singapore, as with many densely inhabited capital cities, can contribute to the transmission of infectious diseases. However, this and the small geography have also been advantageous in terms of the COVID-19 response, given the ability to more efficiently organize and provide services, compared to more sparsely populated regions over which resources are dispersed.

As an international travel hub, Singapore sees a large flow of travelers and goods, through the country which are major contributors to its economy. But the nature of global travel leaves countries susceptible to the importation of infectious diseases. With major travel links with China, Singapore confirmed its first imported COVID-19 case from Wuhan, China on January 23, 2020. The Delta variant became dominant in Singapore by May 2021 [1], after the variant was first detected globally in October 2020, though the number of cases remained low for several months, thanks to stringent containment measures. The first two imported cases of the Omicron variant were detected on December 2, 2021, soon after the variant was first detected globally in November 2021 [2]. With a developed knowledge economy and high rates of digital literacy, the Singaporean population is known to be very amenable to technological solutions. This includes the introduction of SingPass—a digital identity authentication app for making government transactions, and HealthHub—an app and web platform allowing residents to access part of their medical records and their appointments. These existing instances provided part of the foundational digital infrastructure upon which COVID-19-specific technological solutions were built.

The public health system in Singapore

Singapore's inpatient hospital services are largely provided by the public sector, with a smaller proportion provided by the private sector. However, these roles are flipped for acute visits at primary care centers, where services are most often provided by the private sector, with a smaller proportion provided by the public sector.[1]

Pandemic response system

The country's response to the COVID-19 pandemic was in part shaped by its previous experiences with pandemics and local disease outbreaks. These experiences provided the foundation for Singapore's pandemic response frameworks and led to investments and the subsequent development of robust pandemic response capabilities.

Singapore had previously dealt with several major infectious disease outbreaks before the COVID-19 pandemic. These include the severe acute respiratory syndrome (SARS) epidemic in 2003, the H1N1 influenza pandemic in 2009, and a Zika virus outbreak in 2016. The SARS outbreak directly affected Singapore with substantial health and socioeconomic impacts and was a notable event that triggered major developments in the disease response system. One such development was that of the Disease Outbreak Response System Condition framework, which communicates the threat level of the disease and provides a structure to conceptualize and convey necessary public health measures (Fig. 5.1). This framework was continuously improved through reiterations across the years—based on accrued experiences and lessons from responding to subsequent infectious disease threats—and was later used to guide the response to COVID-19. Planning exercises and tabletop exercises have also been used to refine policies and coordination between internal governmental agencies, as well as between regional and international counterparts.

Singapore had also made infrastructural improvements in its capabilities to manage communicable diseases and outbreak threats. After the SARS pandemic, the Communicable Disease Centre was renovated and a new wing was added, which increased its isolation capacity [5]. This center remained the main infectious disease treatment facility until 2019 when it was replaced by the National Centre for Infectious Diseases (NCID)—an interdisciplinary center that incorporates clinical, public health, laboratory, and research expertise. The operational capabilities of the NCID were first tested before

---

1. One notable exception to this trend is chronic disease care, which is most often provided by the public sector.

	Disease Outbreak Response System Condition (DORSCON) Alert Levels			
	Green	**Yellow**	**Orange**	**Red**
Nature of Disease	Low/limited virulence OR High virulence but no/limited human-to-human transmission	High virulence and transmissibility but mainly overseas OR Low virulence but high transmissibility	High virulence and transmissibility with limited local spread that is being contained	High virulence and transmissibility with widespread transmission
Impact	No disruption	Minimal disruption (e.g., border screenings)	Moderate disruption (e.g., quarantine, hospital visitor restriction)	Major disruption (e.g., work from home, school closures)
Public Advice	Be socially responsible and maintain good personal hygiene			
		Look out for health advisories		
			Comply with control measures	
				Social distancing and avoid crowds

FIGURE 5.1 The Disease Outbreak Response System Condition framework adopted by Singapore. This system was initially developed after the SARS epidemic in 2003, further refined thereafter, and used during the COVID-19 pandemic [3,4].

the COVID-19 pandemic when it responded to an imported mpox[2] case that was detected in May 2019.

Whole-of-government crisis response system

Singapore had previously developed a whole-of-government crisis management structure, designed to deal with the wide-ranging issues and responses that arise in a crisis. The Homefront Crisis Executive Group was formed in 2004 and comprises senior representatives across governmental ministries (Fig. 5.2). It was redeveloped from the Executive Group, which was first set up in the 1970s for responses to security-related issues. For the response to COVID-19, at the level of political leadership, the Homefront Crisis Executive Group worked with a Multi-Ministry Task Force (i.e., in place of the Homefront Crisis Ministerial Committee) that was announced on January 22, 2020, to respond to the risk of an impending pandemic. The structure enabled coordination across government agencies and the mobilization of resources necessary for pandemic response and service provision.

With these, Singapore had structures and processes in place to prepare for and respond to pandemics. The 2018 Joint External Evaluation of

---

2. Until late 2022, Mpox was known as monkeypox following a tradition in which the names of poxviruses and poxvirus-related diseases were based on the animal in which the disease was first described.

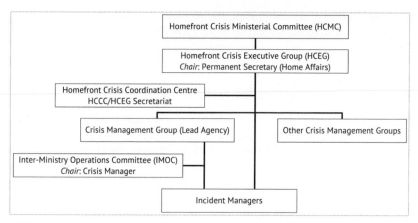

FIGURE 5.2   Structure of Singapore's Homefront Crisis Management System [5].

International Health Regulation (IHR) core capacities of Singapore by the World Health Organization noted high levels of achievement in fulfilling IHR requirements in Singapore, including strong national commitment, whole-of-government mechanisms, scalable responses, and use of innovation and new technologies [6].

Legal public health preparedness

The Infectious Diseases Act (IDA) grants powers to the Minister for Health and Director of Medical Services (the chief medical officer, redesignated as the Director-General of Health in June 2023) at the Ministry of Health to carry out public health actions required to control the spread of infectious diseases in Singapore [7]. It allows for the imposition of legal requirements on persons to undergo diagnostic tests, quarantine, and isolation, as well as medical treatment for infectious diseases of concern. The IDA has been strengthened over time by the need to control endemic diseases in Singapore, such as tuberculosis, as well as threats posed by emerging infectious diseases.

The response measures required during the COVID-19 pandemic were further empowered through the COVID-19 Temporary Measures Act [8]. Aside from measures that provided relief to those affected by COVID-19 (e.g., delays in contracted work), this Act provided additional legal backing for the public health actions required for COVID-19, including community measures such as activity restrictions, mask-wearing, the conduct of social and economic activity in a safe manner, and vaccination requirements. Ultimately, the two Acts enabled the effective deployment of COVID-19 measures. Subsidiary legislation under the two Acts was updated as the COVID-19 responses evolved and represented an important lever for ensuring the adoption of public health measures throughout the pandemic response. The swiftness of passing the Act and the nimbleness by which the legislation was updated according to the COVID-19 situation were critical to enabling the prompt and effective public health response.

Surveillance and testing for COVID-19

Until late 2021, Singapore's COVID-19 response comprised disease containment strategies, particularly in the initial response where vaccines were unavailable. As with any infectious disease outbreak, the crucial first step in disease containment is the ability to detect cases through testing and surveillance capabilities. These capabilities are also fundamental for understanding other biological, medical, and epidemiological aspects of a given pathogen and the diseases caused by it.

COVID-19 was designated as a notifiable disease at the start of the outbreak. Accordingly, hospitals in Singapore were required to submit daily reports on their COVID-19 patients, while laboratories were required to report on the results of all SARS-CoV-2 tests performed. As the pandemic progressed, information and technology systems were developed to ensure the flow of COVID-19-related clinical and laboratory information to national systems, and to provide a comprehensive picture of the situation.

Genomic surveillance was performed through whole genome sequencing of cases at the National Public Health Laboratory. Phylogenetic analysis was also performed to aid epidemiological investigations and to map out the relationship between clusters. In the initial stages of the pandemic, when case numbers were low, all specimens with sufficient viral loads were sequenced. However, as the pandemic progressed, sequencing shifted toward a purposive sampling approach, with specific case types, such as severe or imported cases, prioritized for sequencing.

Evolution of SARS-CoV-2 testing modalities and their adoption

Diagnostic tools are needed to detect cases of disease. At the start of January 2020, there was limited information on SARS-CoV-2. Instead of a specific test, a pan-coronavirus polymerase chain reaction (PCR) assay was initially used. By mid-January 2020, when genetic sequence information on the SARS-CoV-2 became available, a more specific PCR test was developed by the National Public Health Laboratory.

While PCR testing was, and remains, the gold standard for SARS-CoV-2 diagnostic testing, Singapore also sought additional alternatives. Academic partners were invaluable in pursuing these goals. For instance, serology tests were quickly developed by local researchers, and a viral neutralization assay and an ELISA assay for SARS-CoV-2 were employed in February 2020 for epidemiological investigations into early clusters of COVID-19 [9]. The development of serology tests also led to the eventual development of the cPass™ SARS-CoV-2 neutralization antibody detection kit [10].

Singapore also actively investigated methods of specimen sampling for PCR testing and characterized the test performance of various sampling methods including nasopharyngeal, mid-turbinate/oropharyngeal, anterior nasal, deep-throat saliva, and pooled sampling. Pooled sampling methods proved to be particularly useful, as they could extend testing capacity to cover large numbers in a population where infection rates (and test-positivity rates) were still low—as was the case in Singapore until the Delta wave arrived in August 2021.

More convenient tests were also pursued and investigated. For instance, breathalyzers were investigated as one alternate form of diagnostic test. While convenient, breathalyzer tests were ultimately unable to meet the requirements for the detection of the SARS-CoV-2 virus. The development and validation of sufficiently sensitive and specific antigen rapid test (ART) kits, however, was a significant breakthrough and these tests rapidly replaced PCR as the main testing method for the general population in the later phases of the pandemic, when the epidemiological situation meant that efficiency of testing large numbers of people was more important than the additional sensitivity provided by the PCR test, with PCR tests reserved for certain clinical management settings.[3]

Testing of target populations

Testing of symptomatic persons

In the initial months of the COVID-19 pandemic, Singapore's testing approach was focused on persons who fulfilled suspect case definitions, which included clinical and epidemiological criteria. Generally, these were people who had traveled to countries in which there was local transmission of COVID-19 or were close contacts of other confirmed cases. Suspect cases had to undergo PCR testing at hospitals, which was legally enforced under the Infectious Diseases Act where needed. Testing for all pneumonia cases at hospitals was also done, regardless of contact or travel history, to pick up unlinked cases.

However, as transmission became established locally in Singapore and more unlinked cases emerged, there was a need to test more widely. Health authorities determined that testing should be extended to all symptomatic acute respiratory infection (ARI) cases. ARIs are very common, and accordingly, testing all ARI cases required substantial PCR testing capacity. To meet this demand, the government tapped private laboratories to perform SARS-CoV-2 PCR tests for national COVID-19 testing and made capital

---

3. This shift occurred in tandem with a change in the country's approach toward COVID-19 and is described in greater detail later in this chapter.

investments to strengthen PCR testing capacities in both private and public laboratories. Testing for ARI and suspect cases then shifted from the hospital setting to primary care clinics. Patients presenting with ARI to medical providers were strongly encouraged to undergo PCR testing and were otherwise issued a 5-day legal order to remain at home if they declined.

ARI cases were generally served by medical practitioners, who would be reimbursed by the government for performing the swabs. This initially was done through public sector polyclinics and Public Health Preparedness Clinics (PHPCs)—the latter of which are private primary care clinics that, prior to COVID-19, were enlisted in a scheme under which they agreed to contribute toward specific public health functions during emergencies (e.g., administering vaccinations or dispensing of therapeutics). ARI testing was subsequently expanded more widely across the primary care sector to perform swabs for symptomatic patients.

Testing of asymptomatic persons: rostered routine testing and expanding PCR testing capacity

In April 2020, once clear evidence had emerged supporting the possibility of presymptomatic and asymptomatic COVID-19 transmission, testing approaches in Singapore expanded to include the testing of asymptomatic persons. As a significant proportion of cases could be without symptoms at a given point in time, Singapore, having adopted a stringent containment approach, extended PCR testing to asymptomatic close contacts and inbound travelers at the start and end of their quarantine periods. The need for testing capacity also increased substantially when Singapore introduced a Rostered Routine Testing (RRT) Program, which initially consisted of fortnightly PCR tests for all dormitory-dwelling migrant workers. This program was implemented from around mid-2020 in response to a large outbreak in this setting that occurred in April 2020 and aimed to detect presymptomatic or asymptomatic cases early to reduce disease transmission that was occurring in this specific, densely populated setting; it eventually ceased on March 22, 2022. Besides expanding PCR testing capacities, pooling multiple specimens from asymptomatic persons in a single PCR reaction helped to increase the number of persons that could be tested with the same number of tests (i.e., up to five specimens in one PCR test).

Local clusters, including the previously referenced dormitory outbreak, demonstrated how the spread of COVID-19 could be exponential and highlighted how early detection was critical for a containment strategy to be successful. RRT, while resource-intensive, was therefore viewed as an important tool to maintain a containment strategy in the face of a pathogen that could transmit while its host was presymptomatic or asymptomatic. Eventually, RRT was extended to settings determined to be at higher risk of infection and to those living or working in settings with persons who had a greater vulnerability to COVID-19. This included border workers, frontline workers who interface with potentially infected persons, nursing home residents and staff, and health care workers.

Initially, RRT was performed at the sites where tests were required, with special attention given to dormitories and at quarantine facilities. However, as RRT was extended to more work settings as a requirement, there was a clear need to consolidate testing capacity and establish additional testing centers. Further, it was not possible for the Ministry of Health alone to monitor the RRT requirements for such a wide range of workers and to organize tests for these individuals. To address this, specific workplaces were organized by sector, and public sector agencies were appointed as sector leads; these agencies were then responsible for ensuring that testing requirements were met.

In addition to the RRT programs, the testing of inbound travelers and close contacts, which was required as a part of quarantine requirements, also contributed to testing demand. It was determined early on that additional testing capacity, beyond that provided by the formal healthcare system, would be required. Given both the scale of operations and the need for oversight over the national testing capacity, a Testing Operations Task Group was established, to oversee these efforts.

However, the widespread PCR testing of symptomatic ARI cases, RRT programs, and travelers did not only require substantial laboratory capacity but also workforce capacity to perform swabs for specimen collection. This initially proved challenging, given that the health care workforce was already stretched thin by the medical response required for providing care to infected persons. To address this, Singapore embarked on a recruitment exercise for swabbers from the general population—most often individuals without a background in health care. This approach proved feasible and a practical option for remedying this COVID-19 manpower challenge, as many jobs had been displaced by the secondary effects of the pandemic. These individuals were trained and supervised by formal health care workers, most often nurses, and enabled Singapore to achieve its goals of increasing testing capacity. With these concerted efforts to expand testing, PCR testing capacity was progressively increased to 60,000 tests per day within a year.

Testing of asymptomatic persons: extended testing with antigen rapid tests and wastewater surveillance

Toward the end of 2020, ARTs emerged as a relatively convenient testing solution. While the test performance was inferior to that offered by PCR testing, ARTs also presented many advantages including that they did not need to be done in a laboratory, provided fast results, and were relatively cheap. Because of these considerations, ARTs were deemed to be more scalable for repeat testing at the population level, when compared to PCR.

This allowed for new testing strategies, such as pre-event testing and daily testing of close contacts. Pre-event testing was introduced in Singapore in early 2021 to allow for higher-risk activities, such as mass events or visiting nursing homes, while reducing the risk of an infected person attending the event. Initially, ARTs were still required to be administered by licensed providers for such a purpose and were provided by private providers. Another example of ART-facilitated testing strategies includes the use of ARTs to augment RRT programs. ARTs were added to testing schedules in higher-risk settings, most often between the fortnightly PCR tests. As more data emerged around self-administered ARTs, the modality of testing shifted from a provider-administered swab to one that relied on self-performed tests, although Singapore required supervision of self-performed tests for the result to be officially admissible (i.e., through direct or video observation). The test used for RRT also shifted from PCR tests to ARTs.

With the emergence of the Delta variant in August 2021 and a marked surge in case counts, ART self-testing provided a feasible solution for population-wide testing, in the context of a well-vaccinated population. From August 29 to September 27, 2021, Singapore completed its first nationwide distribution of ART kits, providing each household with six tests. These kits also provided an accessible testing method for the increasing number of cases and their close contacts, who were then placed in home isolation and home quarantine. Additional ART self-testing kits were made available to close contacts through vending machines deployed across the country.

Another major shift toward ART testing came with revisions to Singapore's COVID-19 protocols on October 9, 2021, as the strategy shifted from one prioritizing containment to one that prioritized reducing the spread of disease and avoiding overwhelming the health care system. As a result, PCR testing was no longer required to confirm the diagnosis of COVID-19 in the community unless individuals were medically vulnerable or had to be admitted for care. This drastically reduced the demand for PCR testing and ARTs became the primary diagnostic tool for the diagnosis of COVID-19.

Acknowledging that there may be a lag or even a reduction in health-seeking and clinical testing, wastewater testing was also deployed with the intent of providing an early and supplementary indication of disease activity in the community. It provided an additional measure of disease activity in the population, the results of which would not be affected by changes in SARS-CoV-2 testing rates over time.

In summary, Singapore's testing strategy was characterized by extensive testing in the initial phase, which later shifted toward ART self-testing from August 2021 onward as the strategy shifted toward living with COVID-19 (Table 5.1). The initial PCR-based testing strategy was a mammoth task—requiring large amounts of resources and rapid capacity strengthening. However, the demand for PCR testing gradually decreased as other forms of testing became available that have helped to provide more pervasive testing and allowed for cross-checking of surveillance signals—ultimately enabling decision-making in the rapidly evolving pandemic.

TABLE 5.1 SARS-CoV-2 testing programs employed in Singapore.

Testing program	Description and remarks
PCR testing in hospitals for patients who are suspect cases	Targeted surveillance of cases requiring hospital care was commenced early in the pandemic. This evolved to more general testing of patients with acute respiratory infections for infection control and surveillance, with an eventual transition toward the use of antigen rapid tests.
Testing of close contacts	Close contacts identified through contact tracing underwent PCR testing, which was a part of quarantine requirements. Active contact tracing in the community was discontinued in late 2021.
"Swab and Send Home" for persons with respiratory symptoms fulfilling the suspect case definitions in the community	The "Swab and Send Home" program in primary care was initially offered through Public Health Preparedness Clinics (PHPCs). First limited to testing of suspect cases, it was later extended to include testing of all persons with acute respiratory infections in the community. The demand for testing required a substantial testing capacity, and aside from increasing laboratory capacity, it called on testing providers beyond the PHPCs to set up testing centers. These programs were initially PCR-based but transitioned to concurrent PCR and antigen rapid test (ART) testing when ART became available as a testing modality in early 2021. Later that year, only ART tests were needed for those who had low medical risks.
Rostered Routine Testing (RRT) for asymptomatic persons in identified higher-risk subpopulations, which was conducted at regular intervals	Higher-risk subpopulations, such as workers living in high-density dormitories, were required to undergo testing at regular intervals for eligibility to leave their residence for work. This was tracked and reflected on digital records, such as the *SGWorkPass* smartphone app displaying an *AccessCode* status. RRT was PCR-based initially, subsequently involving alternating ART and PCR tests in some subpopulations, before transiting to only requiring ART. The regular testing helped to detect infections early and to curb the spread of COVID-19 but was resource intensive and discontinued in early 2022.
Predeparture and on-arrival testing, and in-quarantine testing for inbound travelers	These testing programs sought to avoid the importation and spread of infections from inbound travelers and evolved throughout the course of the pandemic. Predeparture PCR tests were initially to be conducted while in the origin country, but remotely supervised ART self-testing was eventually allowed. On-arrival tests were conducted at the airport and required establishing testing facilities on-site that were able to meet the testing requirements. Testing of travelers serving their quarantine periods was done through testing conducted on-site at quarantine locations or testing centers.

(Continued)

TABLE 5.1 (Continued)

Testing program	Description and remarks
Testing for discharge of COVID-19 cases	De-isolation of COVID-19 cases relied heavily on PCR test results, requiring negative tests and then low viral loads through high cycle threshold values. The persistent shedding of the virus, however, occasionally resulted in prolonged isolation periods. With additional data on the infectious period, de-isolation requirements evolved to being based on the time elapsed since the start of infection or the achievement of a negative ART self-test.
Pre-event testing for higher-risk events	As a means of preventing transmission while resuming societal activities during the pandemic, pre-event testing was made a requirement for certain events beginning in early 2021. This program required an ART to be performed 24 hours before the given event. Initially, tests had to be administered by a provider initially, but supervised self-swabs, either on-site or virtually online, were eventually accepted.
Wastewater testing	As disease transmission became more widespread in the community, wastewater testing was conducted at a variety of sampling sites as an indicator of disease activity in the population and specific subpopulations.

Isolation and medical management of SARS-CoV-2 infections

Singapore undertook stringent isolation measures toward SARS-CoV-2 infections up until late 2021—adapting its approach based on an evolving understanding of the disease and the practical constraints in doing so. Over time, this approach has evolved from one that was primarily hospital-based to one that is primarily home-based. Health care resources, particularly manpower and hospital beds, were severely limited amid the high burden posed by COVID-19. However, while the health care system faced immense pressure, it managed to reorganize care and alleviate the situation, which enabled it to avoid collapsing.

Singapore's approach to isolation

Initially, when there were limited understanding of the SARS-CoV-2 virus and COVID-19, both suspected and confirmed cases were isolated in negative pressure rooms in hospitals, in particular the NCID. Cases were discharged from the hospital only if they tested PCR-negative twice, consecutively, 24 hours apart. This resulted in lengthy admissions due to what is now recognized as non-infectious viral shedding of the SARS-CoV-2 virus. While the length of admissions was not ideal, this approach promoted a high level of containment and enabled physicians to observe and study patients across their course of illness, allowing them to characterize the clinical, virological, and epidemiological characteristics of the disease.

However, such an approach was resource-intensive and not feasible as case numbers continued to rise. In this context, and based on emerging data on infectiousness that had begun to accrue, Singapore gradually changed its approach to isolation. More specifically, the criteria for discharge changed from requiring negative PCR testing to those that incorporated cycle threshold values and time-limited isolation periods (Table 5.2). There was also a shift from hospital-based isolation to isolation in nonhospital facilities—arising from the observation that a majority of COVID-19 patients were asymptomatic or had mild disease and did not require the level of inpatient care afforded by hospitals.

TABLE 5.2 Brief description of key changes in deisolation criteria for confirmed COVID-19 cases in Singapore[a].

Date of criteria	Deisolation criteria
Start of pandemic	Two consecutive negative PCR tests, spaced $\geq$ 24 hours apart. *As a more stringent criterion, cases in subsequent periods who met it continued to be eligible for deisolation.*
May 2020 [11]	Day 21 of infection with an additional 7 days of medical leave
May 2021 [12]	1. Day 14 or beyond, when the case tests negative on a single PCR test; or 2. Day 21 or beyond, when a case that continues to test positive on PCR achieves two consecutive PCR tests (at least one day apart) with a cycle threshold value (Ct) $\geq$ 35 *In both scenarios, they were discharged with an additional 7 days of medical leave where there was no further need for hospital care.*
Sep 2021 [13]	1. If vaccinated, deisolate on day 10, or, from day six onwards upon achieving a PCR test with Ct $\geq$ 25 (including a negative test); or 2. If unvaccinated, deisolate on Day 21, or from Day 14 upon achieving a PCR test with Ct $\geq$ 25 (including a negative test)
Oct 2021 [14]	1. If vaccinated, deisolate on day 10; or 2. If unvaccinated, deisolate day 14; or 3. After 72 hours of testing positive, when the case tests negative on a SARS-CoV-2 rapid antigen test *The duration specified in (a) and (b) were eventually reduced.*

[a]*There were variations in criteria for specific patient groups, such as that for immunocompromised persons and, when the population was largely naive to SARS-CoV-2 infection, persons who had serological evidence that they were at a late phase of infection.*

Around April 2020, a large outbreak occurred in several migrant worker dormitories. It was also around this time that stringent social and community measures were instituted via a "circuit breaker" (see also the section on Social Measures) to bring cases down to a manageable level in the absence of a vaccine. This outbreak resulted in many healthy working-age men being infected, though the rates of severe disease and deaths were very low. Nevertheless, the need to isolate a large number of cases as a means of halting transmission accelerated the establishment of nonhospital isolation facilities, or Community Care Facilities (CCFs) (Table 5.3). CCFs included temporary beds set up in converted facilities—such as exhibition halls, hotels, and chalets—which were largely vacant due to the COVID-19 travel and activity restrictions. The establishment of these facilities allowed for hospitals to be reserved for vulnerable individuals and sicker patients that required more intensive medical care.

This approach continued with some adjustments until the third quarter of 2021 when the Delta variant resulted in a large surge of infections. It threatened to overwhelm the existing two-tiered system and adjustments were made accordingly. These included establishing Community Treatment Facilities (CTFs), which were converted intermediate care facilities such as community hospitals, set up with limited medical manpower but not as much as that in acute hospitals. CTFs were able to manage COVID-19 patients that were in stable condition but required supplemental oxygen. At this time, there was also a shift from facility-based isolation to home-based isolation for mild cases.

TABLE 5.3 Types of COVID-19 care facilities used in Singapore to respond to the pandemic.

Type of facility	Description
National Centre for Infectious Diseases	National specialty center for infectious diseases that functioned as the main receiving center for COVID-19 cases.
Hospitals	Public sector hospitals managed cases of COVID-19 that presented directly to them.
Community treatment and community care facilities [15]	Community facilities that were established to provide for patients requiring lower acuity of care, thereby preserving hospital capacity. It particularly catered to at-risk but stable patients requiring closer monitoring such as elderly patients and those with other comorbidities.
Community isolation facilities	Community facilities without health care manning, for well (i.e., asymptomatic) or mild COVID-19 cases.

Only persons who were unable to isolate in their own homes were isolated in facilities (e.g., those who cohabitated with medically vulnerable or elderly individuals). Finally, there was a shift in the provision of care toward an approach that emphasized primary care physicians and telemedicine. Primary care physicians were asked to help triage and differentiate patients that required hospital/CTF care and those that could be managed at the primary care level, with telemedicine support when needed. This shift allowed for vulnerable but well patients to be monitored in the community, without needing to be admitted to hospitals. Support was provided to these patients, including oximeters, and daily phone checks to ensure they remained well.

Isolating infected cases at home was a major shift in Singapore's COVID-19 response. This home-based approach for COVID-19 isolation was adopted for the remainder of the local pandemic response, although no legal orders were imposed, unlike the case for other types of isolation. And, as previously referenced, this change was also accompanied by a change in the isolation discharge criteria that emphasized an ART-guided discharge or a time-limited 7-day isolation for vaccinated persons. These revisions were triggered by a need to address the surging case counts, but also a reality that the population had attained high levels of vaccination coverage, resulting in substantial amounts of vaccine protection against severe disease.[4]

The medical response to COVID-19

The entire health care system was involved in the COVID-19 response. The NCID served as the primary receiving hospital for COVID-19 cases detected in the community, while other public hospitals admitted patients who presented directly to them. This helped to reduce the burden that would otherwise be placed on other public hospitals. As the need for beds grew, the bed capacity of private hospitals was also utilized. Community facilities, such as CCFs and CTFs, interfaced with the hospitals as they were frequently the discharge destination for well or stable patients who were still infectious and required isolation. The medical treatment for all COVID-19 cases was fully subsidized by the government, except for short-term visitors to the country, who were required to have medical insurance.

The Ministry of Health (MOH) monitored the health care capacity across the country and held regular meetings with hospital leaders and clinicians to discuss the clinical, public health, and health service aspects of the pandemic response. Adjustments in policies were made in consultation with key clinical and hospital leaders. The MOH also coordinated the flow of patients through its crisis operations arm—identifying them from diagnosis to admission, isolation, and discharge.

---

4. It was estimated that 76% of the population had received the COVID-19 vaccine by August 1, 2020. Vaccination in Singapore is discussed in greater detail later in this chapter.

Overall, Singapore's response was tightly linked to data and evidence, and the same holds true for the medical response. The NCID was involved in bringing together public hospitals in Singapore for key studies on COVID-19 cases, conducting clinical studies, and participating in multicenter drug trials. Beyond these efforts and the invaluable data provided by them, the NCID also provided opportunities to access COVID-19 therapeutics, some of which were still in development.

Throughout the pandemic, health care resources and equipment were in limited supply, and Singapore consciously and aggressively established a resilient supply and stockpile by engaging a wide range of partners in the supply chain. One advantage for Singapore is that due to its relatively smaller size, the supplies required were generally less compared to the demands of other countries, making it easier for suppliers to fulfill their requests.

Still, not all resource shortages are this straightforward. Healthcare manpower, for instance, was a major constraint and difficult to rapidly expand during the COVID-19 response. With the surge in healthcare demand due to COVID-19, a re-prioritization of resources was necessary. Nonurgent and noncritical medical care procedures were delayed during surges of COVID-19 waves to better manage the COVID-19 cases both within hospitals and outside of them with re-distribution of medical manpower. Additionally, public and private sector medical providers were recruited to run the CTFs. Finally, to further mobilize healthcare workers who wanted to contribute but were not already involved in the COVID-19 response, a volunteer pool, the SG Healthcare Corp, was established.

Ultimately, while under a heavy load, the healthcare system of Singapore was not overwhelmed by the COVID-19 pandemic and was able to ensure that essential care continued to be provided. Healthcare workers formed the backbone of the response to COVID-19, and it was through their perseverance and sacrifices that the country was able to weather the brunt of the pandemic. If there are any key lessons to be learned from the COVID-19 pandemic, one is certainly the importance of supporting frontline workers in times of crisis, as they are essential to the response.

Contact tracing and quarantine

For the period that Singapore maintained a disease containment approach, contact tracing and quarantine were key public health strategies. This involved comprehensive and resource-intensive efforts that were sustained until August 2021 (i.e., the emergence and surge caused by the Delta variant), at which point there was a shift toward depending more on a digital technology-driven approach. The latter provided a sustainable solution when infections in the community were prevalent. Contact tracing in the general community eventually ceased in April 2022 after the Omicron BA.1/2 variant wave subsided (Table 5.4).

TABLE 5.4 Evolution of the contact tracing and quarantine measures adopted in Singapore.

Time period	Contact tracing	Quarantine or quarantine-like measures
Start of pandemic—Aug 2021	Manual contact tracing through phone interviews of cases. Subsequently augmented with proximity logging technology (TraceTogether app or token) from September 10, 2020.	14-day quarantine in a dedicated quarantine facility, including the use of holiday accommodations, such as hotels.
Aug 2021—Oct 2021	Similar to previous approach, but with criteria adjusted for efforts to focus on highest-risk close contacts.	Home quarantine, initially a 14-day period then reduced to a 7-day period. Home quarantine was adopted amid the Delta wave to cater to the large numbers of close contacts arising from the surge in cases.
Nov 2021—Apr 2022	Manual contact tracing ceased in most settings (i.e., those that were not high-risk), and proximity logging technology was used to identify close contacts, who were then notified via SMS. Contact tracing and use of proximity logging technology ceased in April 2022.	Self-testing without quarantine. Instead of a quarantine order, the close contacts were advised to self-test with ARTs and only to continue with daily activities if they tested negative.

At the start of the pandemic, epidemiology teams at hospitals manually traced contacts of their COVID-19 patients jointly with the MOH. However, as case counts rose and with a sustained approach of containment, more resources were required. To address this gap, a dedicated COVID-19 contact tracing team was set up at the MOH, by recruiting manpower from across the ministry, including officers who had previously had relevant experience in communicable diseases.

Still, as the number of cases increased, even this initial capacity was exceeded. Given that this occurred during the second quarter of 2020 when effective vaccines and therapeutics were still unavailable, a decision was made to persist with a containment strategy. Accordingly, there was a mobilization of manpower across the wider public sector toward the COVID-19 response. The Singapore Armed Forces augmented the capacity for contact tracing by providing large numbers of personnel that were used to establish several more contact tracing teams. Further, due to the societal disruptions caused by the COVID-19 pandemic, there were a significant number of individuals in affected industries, such as the airline industry, who were also employed as contact tracers. These individuals were actively recruited to help in the response, which was beneficial for both the pandemic response and the individuals whose jobs were impacted by the pandemic, and similar win-win arrangements could be sought out during the response to future health care crises.

Using the criterion of being within 2 m of an infected person for at least 15 minutes, on average, contact tracing in the community yielded approximately 20 close contacts per COVID-19 case. This included contact with an infected person prior to diagnosis or the development of symptoms, if any, to account for the risks of presymptomatic or asymptomatic transmission.

Close contacts who were identified were quarantined. Given the ratio of cases to close contacts identified, many persons were quarantined. As with the approach for isolation facilities, dedicated quarantine facilities were established, oftentimes in hotels. The duration of the quarantine was initially 14 days in length and generally done individually. However, cohorted quarantine was adopted for family units for which all members were close contacts, as well as for close contacts in migrant worker dormitories as part of the large outbreak in that setting in 2020, where the sheer size of the outbreak made it not possible to individually quarantine the contacts.

Contact tracing is a manpower-intensive response activity and to harness technology to ease these manpower demands, Singapore developed a proximity logging Bluetooth solution, dubbed "TraceTogether," and adopted it nationwide in September 2020. Persons could install the app on their smartphones or could obtain a stand-alone token version. The latter was helpful for those without such devices, such as young children and the elderly. The app also had the functionality of logging entry and exit into venues, known as "SafeEntry" which was made a requirement across major venues. The app was also linked to vaccination status, on which eligibility for entry to certain venues in the mitigation phase was based. By making it a requirement for entering major venues, this led to a widespread adoption of the TraceTogether app or token. This allowed digital logging of locations visited and close contacts exposed, which reduced the heavy reliance on individual recall of these details previously which was time-consuming and prone to recall errors.

For all of its advantages, proximity logging raised notable privacy concerns, and the government assured the public that the data gathered would only be used for COVID-19 contact tracing purposes. The proximity data was logged locally on the person's device and was not transmitted elsewhere unless determined to be a COVID-19 case; only then, would the person need to allow the data to be accessed and shared with the contact tracing teams. To further address privacy concerns over its use, a bill to specifically address the use of the data collected was passed on February 2, 2021 and stipulated that the data could only be used for serious crimes [16]. This demonstrated the importance of stringently and promptly addressing privacy concerns in the implementation of public health measures to gain and maintain public confidence and trust.

This approach for contact tracing and quarantine was largely maintained, until the major surge in cases arising from the Delta in August 2021. As mentioned previously, high vaccine coverage rates were attained by this time, and the high number of cases and resulting close contacts provided the impetus for shifting away from a strict containment approach. With the experiences gained working with the proximity data, the approach shifted toward relying on this alone, and manual contact tracing efforts involving phone interviews with COVID-19 cases were decreased, initially targeting those at highest risk (e.g., family members) before being largely discontinued. As a result, the process of identifying close contacts and notifying them was now automated and close contacts were advised to exercise precaution and quarantine themselves at home instead through a "Health Risk Notice" issued via phone SMS—measures that relied predominantly on social responsibility. As the Delta wave progressed, the advice to self-quarantine in the event of exposure was later revised to recommend daily self-test with an ART for a period of 7 days before individuals continued with their daily activities if they were negative and asymptomatic.

These major shifts to the approach used for contact tracing and quarantine drastically reduced the manpower required for these activities. Nevertheless, manual contact tracing was still offered in selected high-risk settings, such as nursing homes, to comprehensively contain outbreaks. This mode of identifying contacts and issuing advice continued until April 26, 2022, after the Omicron BA.1/2 wave of infection, when the "Health Risk Notices" were discontinued.

Vaccination

Vaccination was deemed essential for "living with COVID-19." Early in the pandemic, vaccines held the promise of preventing transmission in addition to preventing severe disease. As with many other countries, Singapore pursued vaccination aggressively and took measures to ensure that the population could and would achieve high vaccination coverage rates. Working toward this goal involved various efforts and approaches that differed according to the availability of vaccines.

In early 2020, when COVID-19 vaccine development efforts were nascent, the government set up a Therapeutics and Vaccination Planning Group to review promising vaccine candidates and take steps necessary for securing a supply of vaccines. At that time, there was no guarantee that a safe and effective vaccine would even be possible, and vaccine data were still being evaluated. Yet, while weighing the risks associated with missing the opportunity to vaccinate the population, in mid-2020, the government made the decision to arrange advance purchase agreements based on these imperfect data. This action reflected a forward-looking stance, and also one unafraid of measured and calculated risk.

Such an approach soon paid off, as several of the Therapeutics and Vaccination Planning Group's preferred choices, such as the mRNA vaccines developed by Pfizer-BioNTech and Moderna, showed high vaccine effectiveness and were validated through clinical trials and observational studies. The first vaccine stocks, comprised the Pfizer-BioNTech Comirnaty vaccine, arrived in Singapore on December 21, 2020. The vaccines were quickly rolled out thereafter. An Expert Committee on COVID-19 Vaccination was also formed to advise the government on the use of the COVID-19 vaccines. This group comprised infectious disease specialists, immunologists, as well as experts involved in regulatory approval.

The planning and implementation of the national vaccination program were overseen by the MOH's Crisis Strategy and Operations Group. At the time, based on what was known about the vaccines and given the high levels of vaccine efficacy against infection, the goal for vaccination efforts was herd immunity.[5] As such, Singapore strove to vaccinate as much of the population in the shortest amount of time possible and established vaccination centers located across the island that would be able to vaccinate at high volumes (Fig. 5.3).

The vaccination centers were operated under a model that contracted vaccine service provision to private health care groups. These centers were centrally located in neighborhoods as a means of promoting high levels of community visibility. Furthermore, working with major centers facilitated the logistics of vaccine distribution, particularly as the mRNA vaccines for COVID-19 required storage at very low storage temperatures, not typically catered for by other facilities. Selected primary care providers, namely public sector polyclinics and private clinics that were PHPCs, were also included in the provision of vaccinations. Compared to the vaccination centers, these clinics functioned mainly in a supportive role during the main vaccination drive, as their throughput was relatively low. However, they remained important for longer-term vaccination efforts after the immediate need for rapid population-wide vaccination coverage was achieved.

COVID-19 vaccination services were also provided by mobile and home vaccination teams that sought to improve access to vaccines.

---

5. Herd immunity is defined as a form of indirect protection that is enjoyed by a population when a sufficient percentage of that population has become immune to an infection, whether through previous infection or vaccination.

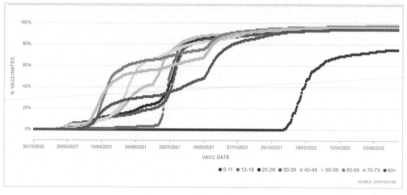

FIGURE 5.3 Primary series vaccination take up by age group (completed full regimen) in Singapore. This shows a greater increase in vaccination coverage after vaccine-differentiated safe management measures were introduced in August 2021, and how Singapore managed to achieve significant vaccination coverage by the time major surges of the Delta and Omicron BA.1/BA.2 wave occurred in Singapore. *Reproduced from data.gov.sg under Singapore Open Data License [17].*

These modes of vaccine provision placed a special emphasis on bolstering vaccine coverage in elderly populations who were more vulnerable and could have mobility issues.

Promoting vaccine uptake

There were several challenges associated with achieving high vaccine coverage including: (1) initially limited stocks of vaccine, (2) the need to address concerns about the new mRNA technology used in vaccines and that they were developed in a relatively short period of time, (3) the need to proceed cautiously in the face of severe, albeit rare, adverse reactions to vaccines (e.g., anaphylaxis, and subsequently myocarditis/pericarditis), and (4) the need to use different models and substantial resources to provide population-wide vaccination. Singapore adopted a combination of approaches to improve the vaccination rates in Singapore.

Given the limited stock of vaccines early in vaccination efforts, vaccination was prioritized for persons at greater risk of severe disease and exposure to the SARS-CoV-2 virus. This initially included individuals aged 70 years and older, as well as frontline COVID-19 workers and healthcare workers. Vaccination was then gradually extended to other demographic groups and subpopulations. This approach allowed authorities to better understand both the risks and benefits as data accrued and to adjust the vaccine recommendations, as needed. It also allowed for a progressive ramp-up of vaccination services to cater to the larger segments of the population and helped to regulate the demand for the vaccine. Still, given the rapidly evolving knowledge on vaccination and the subsequent impacts on recommendations, fluctuations in demand presented a challenge that required maintaining buffers, ramp-up plans, and the judicious use of resources.

The MOH's vaccination program also provided recommended vaccines free of charge to all Singapore residents, including migrant workers and other long-term pass holders. Certain vaccines were also made available via private provision at a cost to individuals. Efforts to increase access to vaccines, such as the mobile and home vaccination teams described earlier, also helped to facilitate the uptake of vaccines by making vaccination convenient to the population. SMS reminders were also used to encourage persons who were eligible for vaccination to register and be vaccinated. An online registration system allowed vaccine-eligible persons to easily schedule an appointment, and walk-in appointments were gradually allowed as demand for the vaccine eased, with priority given to seniors.

Public communications and engagement were also paramount for updating the public on vaccine recommendations and building confidence in the vaccines. In Singapore, these efforts included advertisements on mass media platforms, as well as a social media engagement strategy. Webinars and physical engagement sessions were also organized with grassroots leaders, religious organizations, and unions, among others to effectively engage the community and promote uptake.

It was also important to keep medical professionals updated as they play a crucial role in advising patients and encouraging vaccination uptake. MOH issued professional circulars and in partnership with professional organizations held webinars as well as online town hall sessions to keep medical professionals updated on the details of the vaccine recommendations and the medical and public health consideration behind them. Medical specialty colleges and chapters within the Academy of Medicine, Singapore, also developed consensus statements to guide doctors on specific clinical concerns and patient groups.

As was the case in many other countries, anti-vaccination sentiments were a challenge for COVID-19 vaccination campaigns in Singapore. Prior to the COVID-19 pandemic, and when considering vaccination efforts more broadly, Singapore had achieved high childhood vaccination coverage. Adult vaccinations (e.g., those for seasonal influenza) lagged behind though not primarily due to anti-vaccination sentiments. Still, for COVID-19 vaccines specifically, there were vocal groups that also tapped on anti-vaccination opinions expressed both domestically, as well as overseas.

This highlighted the importance to build and maintain confidence in the COVID-19 vaccine and to address misinformation that might undermine this. Surveys were conducted to better understand the hesitant sentiments being expressed in communities, which helped to target communication efforts. But, given the public health implications, the government also took a more strident stance against this. There were several misleading claims made that discredited COVID-19 vaccines. Where these false statements were made online, legal action was taken under the 2019 Protection from Online Falsehoods and Manipulation Act, to require the publisher of these sentiments to (1) state that the statements were untrue and (2) provide a link redirecting readers to a corrective statement issued by the government. This is akin to a right of reply, adapted to the online space. As a result of all of the efforts previously described, COVID-19 vaccine hesitancy in Singapore stood at 9.9% 6 months after the launch of the adult COVID-19 vaccination program—low, relative to other countries with comparable contexts [18].

Considering the above, Singapore has been successful in achieving high levels of vaccination coverage (Fig. 5.3). The introduction of vaccine-differentiated safe management measures in August 2021 further increased this. Under these measures, only vaccinated individuals were allowed to enter nonessential public venues, such as shopping malls, and allowed to physically return to workplaces.[6] This decision was made by determining that fully vaccinated individuals enjoyed a high level of protection against COVID-19, and particularly severe disease, when compared to unvaccinated persons. The vaccine-differentiated measures, therefore, enabled previously curtailed activities to resume in a safe manner as community, economic, and social activities gradually reopened.

Social measures

After community transmission of the SARS-CoV-2 virus was established in Singapore around February 2020—a reality that was proven by the emergence of unlinked community cases—broader mitigation measures were needed to limit the spread of the disease within local communities. These measures were introduced progressively, ultimately culminating in a "circuit breaker" in April 2020, which was comparable to lockdowns adopted in other countries. Thereafter, measures underwent a few cycles of easing and tightening in response to surges in local cases, reflecting "the hammer and the dance" approach [19]. It was between April and May 2022, when the Omicron BA.1/BA.2 surge had subsided, that these measures were significantly relaxed, allowing for a semblance of social normalcy. By this point, many vulnerable persons had received booster doses and a substantial proportion of the population also had additional immunity from infection.

Additional social measures—including the wearing of facemasks, restricting nonessential businesses, limiting the size of events, practicing social/physical distancing, and working from home—were important for limiting the number of COVID-19 infections and cases of severe disease in the early phases of the pandemic (Table 5.5). They also allowed for other response

---

6. Prior to the introduction of vaccination differentiated safe management measures, working from home was stipulated and nonessential workers were not allowed to work at their places of employment.

TABLE 5.5 Key safe management social measures adopted in Singapore in response to the COVID-19 pandemic.

Social measure	Description and remarks
Mask-wearing in general population	Recommended when local evidence established that presymptomatic/asymptomatic transmission occurred. This was briefly made a legal requirement for indoor and outdoor settings and was progressively relaxed (e.g., not required for outdoor settings from April 2022) before it was no longer required from February 2023.
Restrictions to nonessential businesses and services	Ceased during the "circuit-breaker" period and resumed with restrictions thereafter. The activities determined to be the highest risk, such as attending nightclubs, were not allowed to resume until 2022.
Event sizes	When gradually resumed from August 2020, weddings wakes/funerals, and religious gatherings were first allowed with maximum event size restrictions of up to 20–50 persons, subject to other measures such as distancing and mask-wearing. Subsequently, larger numbers and other events were progressively allowed with the requirements of pre-event testing, and eventually, vaccination requirements (see below).
Group sizes and visits	Group sizes were restricted down to two persons, five persons and ten persons variously at different points of the pandemic to reduce the extent of physical interactions within the community.
Work arrangements	Work from home was the default, except for services that needed workers to be physically present, for which, RRT was implemented. As this measure eased, increasing proportions of the workforce (e.g., 50%, 75%, 100%) were allowed to return to the workplace.
Social/physical distancing	Physical distancing of 2m was imposed on the population. This included seating arrangements, such as those at food and beverage establishments. This in turn also thinned out numbers across all settings.
Vaccine-differentiated safe management measures	From August 2021, given the vaccine protection, these measures were applied to relax the measures described above for vaccinated persons, (while maintaining them for the unvaccinated). Vaccine-differentiated measures were maintained, with adjustments made over time, before being largely relaxed in April 2022.

efforts under the stringent public health response strategy to be viable and effective in the community for most of 2020 and 2021. These measures are briefly described below but are not further elaborated on in this chapter, as they are not the primary focus.

Conclusion

Overall, Singapore had multiple strengths in its capacity and capabilities in pandemic preparedness prior to the COVID-19 pandemic. Nevertheless, COVID-19 was an unprecedented crisis that stretched existing plans and preparations and demanded rapid, large-scale responses. As such, Singapore's response to the COVID-19 pandemic was only possible with a whole-of-government approach and an all-of-society collaboration with stakeholders in various communities. And, drawing upon previous experiences managing crises, including that of previous outbreaks and pandemics, the country launched a swift and nimble response that was able to rapidly adapt to the changing situation and developments, and mobilize a large number of different resources across various responses.

Fundamental to this approach were the outbreak response measures of surveillance and testing, contact tracing, isolation and quarantine, and vaccination. These represent core public health functions and services, capacities of which were increased, and methods transformed, to meet the changing COVID-19 situation. As a result of these actions, Singapore has been able to achieve low excess mortality arising from the pandemic, based on the WHO's estimates for up to year 2021. This was contributed to by high vaccination rates of more than 90% having completed the primary series of COVID-19 vaccinations, high SARS-CoV-2 testing rates, and suppressed community transmission during the initial containment response.

Every country has its own unique context and there are, however, some limitations in extrapolating Singapore's experience to other countries. Singapore has similarities to many developed urban cities. However, it is a city-state and unlike larger countries does not face the challenges in coordinating across multiple levels of governance, such as between national, state, and local responses, which would entail many more parties and complex relationships. The geographical size of Singapore is also relatively small, which reduces the challenge of geographical reach. Besides, the government and political leadership have also been stable with strong capabilities developed over a long period of time, which provided greater willingness to invest in preparedness against future outbreaks and allowed it to act strongly with respect to policies and measures.

Looking ahead, the COVID-19 pandemic is still fraught with uncertainty, especially when considering the unrelenting evolution of the virus and potential subsequent emergence of variants of concern that could lead to clinical outcomes and periodic surges of unknown impact. The challenge for

Singapore, as with many countries, is then to identify a sustainable approach to respond to COVID-19. While vaccination and past infection has given the population a significant level of immunity, the danger that a new variant evades this still lurks. Even as Singapore moved toward an "endemic COVID-19 new normal" with the relaxation of COVID-19 community measures in February 2023, it remained vigilant with ongoing surveillance through local efforts and international partnerships, with plans to reactivate measures where required if threatened by new variants of concern [20]. Beyond COVID-19, there may be known and unknown infectious disease threats that may emerge, and the international mpox outbreak in 2022 demonstrates how susceptible the world is, despite the lessons identified from the COVID-19 response.

While future health threats may come in different forms, and the extent of public health responses may vary, the key measures and principles that underlie the responses remain constant. There is a need to continue to build capabilities across domains, and while it is not possible to indefinitely maintain the level of resourcing required to respond to a major threat like COVID-19, this crisis has clearly demonstrated the need for established governmental and societal structures to effectively mobilize to respond to such threats.

References

[1] Tan A. 550 Covid-19 cases infected with Delta variant detected in Singapore so far. The Straits Times 2021; 10 Jun.

[2] Ministry of Health Singapore. Two imported COVID-19 cases tested preliminarily positive for Omicron Variant. Ministry of Health News Highlights 2021; 02 Dec.

[3] Ministry of Health Singapore. (Revised April 2014) Pandemic readiness and response plan for influenza and other acute respiratory diseases. Singapore: Ministry of Health; 2014.

[4] Government of Singapore. What do the different DORSCON levels mean, <https://www.gov.sg/article/what-do-the-different-dorscon-levels-mean>; 2020 [accessed 18.02.23].

[5] Thulaja N.R. Communicable Disease Centre, <https://eresources.nlb.gov.sg/infopedia/articles/SIP_336_2005-01-03.html>; 2017 [accessed 18.02.23].

[6] World Health Organization. Joint external evaluation of IHR core capacities of Singapore: mission report. Geneva: World Health Organization; 2018.

[7] Parliament of Singapore. Infectious diseases act. Singapore: Parliament of Singapore; 1976.

[8] Thirteenth Parliament of Singapore. (No. 103) Parliamentary debates, Vol. 94. Singapore: Parliament of Singapore; 2020.

[9] Yong SEF, Anderson DE, Wei WE, Pang J, Chia WN, Tan CW, et al. Connecting clusters of COVID-19: an epidemiological and serological investigation. Lancet Infectious Diseases 2020;20(7):809−15. Available from: https://doi.org/10.1016/S1473-3099(20)30273-5.

[10] Tan CW, Chia WN, Qin X, Liu P, Chen MIC, Tiu C, et al. A SARS-CoV-2 surrogate virus neutralization test based on antibody-mediated blockage of ACE2−spike

protein−protein interaction. Nature Biotechnology 2020;38(9):1073−8. Available from: https://doi.org/10.1038/s41587-020-0631-z.

[11] Ministry of Health Singapore. Revised discharge criteria for COVID-19 patients. MOH Circular 138/2020. Singapore: Ministry of Health Singapore; 2020.

[12] Ministry of Health Singapore. Updated approach to recovered travellers & revised discharge criteria for COVID-19 positive patients. MOH memo to public and private hospitals. Singapore: Ministry of Health Singapore; 2021.

[13] Ministry of Health. Revised direct CCF-admission/decanting and discharge criteria for COVID-19 positive patients, discharge memos and management of recovered persons who are tested COVID-19 positive. MOH memo to public and private hospitals. Singapore: Ministry of Health Singapore; 2021.

[14] Ministry of Health Singapore. Changes in the approach to COVID-19. Vol. 140/2021, MOH circular. Singapore: Ministry of Health Singapore; 2021.

[15] Ministry of Health Singapore. Setting up of stepped-up community care facilities. Ministry of Health Press Release; 19 Sep 2021.

[16] Fourteenth Parliament of Singapore. COVID-19 (temporary measures) (Amendment) Bill − Singapore Parliamentary Debates, Vol. 95 (No. 17). Singapore; 2021.

[17] Ministry of Health. COVID-19 vaccination data, <https://www.moh.gov.sg/covid-19/vaccination>; 2022 [accessed 10.10.22].

[18] Griva K, Tan KYK, Chan FHF, Periakaruppan R, Ong BWL, Soh ASE, et al. Evaluating rates and determinants of COVID-19 vaccine hesitancy for adults and children in the Singapore population: strengthening our Community's Resilience Against Threats from Emerging Infections (SOCRATEs) Cohort. Vaccines 2021;9(12):1415. Available from: https://doi.org/10.3390/vaccines9121415.

[19] Pueyo T. Coronavirus: the hammer and the dance. Medium 2020;19 Mar.

[20] Ministry of Health Singapore. Singapore to exit acute phase of pandemic <https://www.moh.gov.sg/news-highlights/details/singapore-to-exit-acute-phase-of-pandemic>; 2023 [accessed 12.03.23].

Chapter 6

Future-proofing: Quezon City's response to the COVID-19 pandemic as investments against the next epidemic

Noel Bernardo[1,2], Jaifred Christian Lopez[3], Esperanza Anita E. Arias[4] and Misha Coleman[5,6]

[1]*International SOS, Pasig City, National Capital Region, Philippines, [2]Atlantic Institute, Rhode Trust, Oxford, United Kingdom, [3]Department of Population Health Sciences, Duke University School of Medicine, Durham, NC, United States, [4]Quezon City Health Department, Quezon City, National Capital Region, Philippines, [5]International SOS, Melbourne, VIC, Australia, [6]Department of Population and Global Health, University of Melbourne, Melbourne, VIC, Australia*

Background

Background on Quezon City, Philippines

Quezon City is a highly urbanized city—with a total population of 2,920,068 and a population density of 18,000 individuals per square kilometer in 2020–2021 [1]. The city is divided into six districts and further divided into 142 component barangays.[1] It is a "planned city" that was designed to replace Manila as the capital city of the Philippines due to overcrowding in Manila and the concern that Manila may be an easy target for bombardment by naval guns, given its proximity to a bay [2]. While Quezon City never did replace Manila as the capital, it does represent an important city in the country, as it is the most populous city in the Philippines and makes significant economic contributions to the national economy. The city has never taken on debt nor run a budget deficit and is considered, both domestically and internationally, an economic success story—enabling it to be innovative and to provide national leadership among the other 146 cities located in the Philippines (Fig. 6.1).

---

1. Barangays can be thought of as "villages" within the city.

Inoculating Cities, Volume II: Case Studies of the Urban Response to the COVID-19 Pandemic.
DOI: https://doi.org/10.1016/B978-0-443-18701-8.00018-7

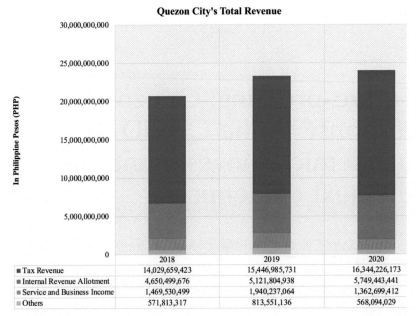

FIGURE 6.1 Annual revenue of Quezon City. *Figure is based on financial report released by the Commission on Audit [1].*

In a city of nearly 3 million people, half of the population is less than 26 years of age and roughly 90% of the population identifies as Roman Catholic [3]. While Quezon City is a prosperous city—with a burgeoning population and a low overall poverty incidence of 1.5%—there are also more than 10,000 families that live below the poverty line [3]. These families and individuals that live in poverty have little or no access to potable water, wastewater sewerage, and sanitation services, which has led the health department to call for the introduction of sentinel and point-of-source water source testing going forward.

At the time of writing, the governance of Quezon City is characterized by strong and progressive leadership—with a female Mayor, the Honorable Mayor Joy Belmonte. Throughout 2020 and 2021, during the emergence and escalation of the COVID-19 pandemic, nearly 40% of the other city councilors were women, as was the head of the Quezon City Health Department (QCHD) whose responsibility was to protect the population [1,4]. The mayor led a highly proactive evidence-based response to the COVID-19 pandemic that had its foundation in the advice provided by the QCHD; this, in contrast with her key political opponent who, among other things, was a prominent advocate for increasing access to and encouraging people to take unproven treatments as a means of fighting COVID-19 (e.g., Ivermectin) [5]. In the most recent elections held in May 2022, the incumbent mayor was re-elected and ultimately earned double the vote of her key opponent, on a two-party-preferred basis.

Quezon City's ability to implement innovative responses and enact policy reforms is hinged on local government autonomy, enshrined in the Philippine Constitution and the Local Government Code of 1991. This devolves significant power to provinces, cities, and municipalities to raise funds, promote investment, and develop and implement policies and programs that respond to their unique local needs. More specific to Quezon City, the solid economic fundamentals and progressive leadership have enabled the introduction of several health and social services that have improved access and enhanced equity. These include, among other initiatives, launching the first central pharmacy where impoverished citizens can regularly avail themselves of free medicines and vitamins; establishing a dedicated city-owned facility to respond to the need for a temporary shelter for abandoned, neglected, and abused senior citizens; operating a residential and rehabilitation center for children who are in conflict with the law; expanding the services of the local public cemetery through the establishment of a public crematorium; creating a Muslim Affairs Service; and promoting healthy motherhood from pregnancy through the first two years of childhood through the Human Milk Bank.

As will be discussed throughout this chapter, Quezon City optimized its local government autonomy and technological environment to implement an innovative response to the COVID-19 pandemic.

Background on the public health system in Quezon City, Philippines

The city's public health system is administered by the QCHD, which is currently directed by an experienced and progressive female public health physician. The QCHD operates three tertiary-level hospitals, 66 barangay health centers, nine "lying-in" clinics, four social hygiene clinics, six animal bite centers, and four sundown clinics [1]. Prior to the COVID-19 pandemic, Quezon City employed the greatest number of public health employees in the region, primarily on a permanent basis and with relatively high levels of remuneration compared to peers in other local governments and health facilities in the National Capital Region. In addition to these public health facilities, there are also 93 privately owned clinics that provide primary health care in the city. Additionally, Quezon City is home to eight hospitals managed by the national government (i.e., financed by the Department of Health), 48 private hospitals, and several more health institutions that are supported by the private sector, nongovernment organizations, and nonprofit corporations.

Still, gaps remain in the local health care system. For instance, to be able to meet universal health coverage requirements of a 1:20,000 facility-to-population ratio, six additional primary care facilities are required in the city.

Further, while all of the previously referenced institutions form an important part of Quezon City's health service delivery network, there is not yet a shared patient data platform or an electronic medical record system in place. Wastewater surveillance capacities are also limited as there are areas that do have a reticulated water supply—severely impacting the city's ability to mount early warning systems and population-wide surveillance.

Meanwhile, the QCHD is also preparing for the full implementation of the Universal Health Care (UHC) Law of 2019. For local governments, these preparations entail forming or strengthening primary care provider networks and province-wide/citywide health systems, increasing focus on primary care services, enhancing infrastructure necessary for health services and data management, and better coordinating with Phil Health, the country's national health insurance system, among other preparatory activities. Prior to the COVID-19 pandemic, QCHD had been working to build its operational capacity to implement the UHC Law through development assistance, as well as investments in data management. These efforts proved helpful as the city eventually bore the brunt of the pandemic.

Additionally, the city was grappling with several communicable disease outbreaks prior to the COVID-19 pandemic. These included African swine fever—a notable outbreak but especially when considering that the Philippines is the world's 10th-largest pork consumer—as well as increased incidence of dengue and tuberculosis cases, a trend which continues at the time of writing [6,7].

The response to COVID-19 in Quezon City, Philippines

Overall, and at the time of writing, the Philippines has had the second-highest number of reported COVID-19 infections in Southeast Asia, trailing Indonesia [8]. It is in this context that Quezon City has implemented an extraordinarily successful response, especially for a city of its size. As of June 30, 2022, since the onset of the pandemic in March 2020, Quezon City has reported a total of 263,954 confirmed cases from hospitals and primary care facilities (roughly 8.9% of its total population), with 261,583 of these individuals or 99.1% of the cases eventually recovering from the illness [9]. Indeed, in the city's 2021 Annual Report, Mayor Joy Belmonte reflected on the city's pandemic response and said, "having more than 1000 average daily cases and an attack rate of nearly 40 per 100,000 would turn other knees to jelly, but our team of front liners were relentless."

Still, it is undeniable that the COVID-19 pandemic presented challenges for Quezon City. For instance, it challenged the city's capacity to deliver much-needed public services, especially as the pandemic disproportionately affected its low-income population. The solid financial foundations of the city government, however, enabled it to embrace innovative approaches for responding to the public health emergency. For example, in recognition of the critical role health care workers played in the fight against COVID-19, the city increased their take-home pay through additional financial incentives—an initiative that also helped the local government to attract and retain qualified personnel to aid in the response. Further, with contractual workers being provided *"plantilla positions,"* the permanency of employment assures the continuity of health operations even after the acute phase of the pandemic response.

As the largest local government unit (LGU) in the National Capital Region, Quezon City's COVID-19 response was designed to align with the available evidence while working to ensure prompt action and coordination with the rest of the Region. More broadly, however, the Quezon City LGU adhered to the National Action Plan Against COVID-19 developed by the National COVID-19 Inter-Agency Task Force. This plan, also known as the PDITR + V Framework, included six pillars: prevention, detection, isolation, treatment, reintegration, and vaccination. The framework provided guidance on the measures necessary for containing, preventing, and eliminating the threats posed by the pandemic and mitigating its social, economic, environmental, and security impacts. It also facilitated the coordination of efforts between the national government agencies, the Quezon City LGU, businesses, private health facilities, civil society, and individuals, thereby ensuring a whole-of-society pandemic response to COVID-19. Thus the response in Quezon City may be characterized as one that was nationally enabled, locally led, and people-centered. The local implementation of this response and the PDITR + V Framework is explained in greater detail herein.

Pillar 1: prevention

Quezon City consistently implemented the various lockdown protocols—locally termed "community quarantines"—that have been promulgated by the National COVID-19 Inter-Agency Task Force since the advent of the pandemic, while also crafting local policies that were more relevant and responsive to the unique needs of its constituents. Examples of these include public safety hours, special concern lockdowns, the promotion of minimum public health standards, and the development of Barangay Health Emergency Response Teams. The public safety hours were a local iteration of nationally recommended curfews. As a part of this preventative intervention, residents in Quezon City were required to shelter in place and local law enforcement was empowered to escort residents to their homes in the event of noncompliance. Special concern lockdowns were a form of stringent monitoring of access to localities with high levels of community transmission. Quezon City's implementation of this measure was unique in that it included the provision of financial aid to households that were impacted by special concern lockdowns, which eventually inspired similar efforts nationwide. Minimum public health standards were promoted through a multisectoral effort that involved the public, businesses, and voluntary sectors in Quezon City. The standards included recommendations for physical distancing, mask-wearing, barrier installations, and provisions for contact tracing. Furthermore, all businesses and offices with more than 10 employees were obliged by local authorities to have response plans to enact should COVID-19 outbreaks among staff occur. Other capacity-building efforts centered on developing Barangay Health Emergency Response Teams. These teams comprised frontline workers that had been trained in patient profiling, contact tracing, patient transport, and other essential response activities (Fig. 6.2). Frontline workers were also provided with a variety of benefits including free testing, personal protective equipment, shuttle services to their places of work, and hazard pay, among others.

Underlying these efforts were those geared toward strengthening local policy and health promotion. On the policy front, the City Council promulgated ordinances and resolutions which empowered implementers, set guidelines for the enforcement of standards, and mandated punitive measures for violations. Meanwhile, on the health promotion front, the Quezon City Public Affairs and Information Services Department developed and implemented a risk communication plan that included messages shared through social media, print media, billboards, and broadcast media, such as the "Why I Wear Masks?" and "#StayAtHome" campaigns. Quezon City also adopted a national health promotion strategy, dubbed *"Bida Solusyon"* which had

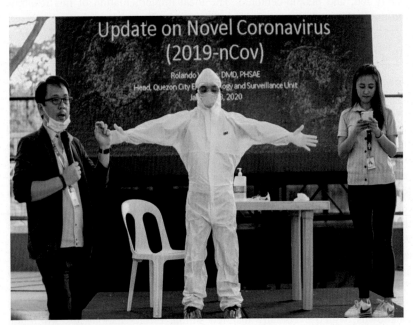

FIGURE 6.2 Training of Barangay Health Emergency Response Teams in Quezon City.

four key components: mask-wearing, hand sanitizing, maintaining a distance, and staying informed.

These health promotion efforts did not escape criticism and opposition, however. For example, when the local government announced that they were the first city in the Philippines to have its COVID-19 Management of the Dead Protocol, one that complied with international standards, and formally announced the opening of its public crematorium facility for COVID-19 mortalities that became functional in March 2020, "netizens" were quick to comment that the mayor was prioritizing cremation over testing, treatment, and tracing efforts. Another example occurred when Quezon City LGU sought to respond to the needs of residents during community quarantine and lockdowns by providing food relief and social amelioration packages. Critics of these actions released fabricated news stories alleging corruption of the city government, despite multiple audit reports recognizing the city for its good governance and transparent financial management practices. The city also faced criticism on several of its social media posts and campaigns.

Likely contributing to this opposition was that many of the response activities were implemented during a politically important time, as it was election campaign season. Nonetheless, improved coordination and efforts between the Public Affairs Department, the Health Office, and other concerned local authorities led to a recalibration of the city's health promotion efforts. This recalibration helped to improve adherence to masking and physical distancing guidelines, and eventually vaccination. These efforts were coupled with the city's social assistance program for low-income families and neighborhoods impacted by lockdowns, which also served as an opportunity to reinforce health promotion efforts.

Meanwhile, empowered by local government autonomy, the city enacted local policies that reduced the nationally mandated penalties for citizens found guilty of failing to wear masks or violating lockdowns. This was primarily to ease the strain of pandemic response measures on low-income and unemployed individuals. These were often the people faced with the tough choice: comply with lockdown regulations, or risk violations in order to seek job opportunities essential for their survival.

Pillar 2: detection

The Detection Pillar of the PDITR + V Framework included efforts to increase diagnostic testing capacity while ensuring equitable access to tests, implementing contact tracing, and conducting data collection and analysis. Cognizant of national-level constraints in testing capacity, Quezon City expanded its local testing capacity by establishing its own molecular laboratory, the Quezon City Molecular Diagnostics Laboratory, with the assistance of various business donors who provided financial and material resources. Initially, this laboratory had the capacity to analyze 1000 samples per day, but this capacity was eventually increased to 4000 samples per day, thanks to additional support from the private sector. Testing services were also made more accessible throughout the city by deploying LGU-operated sample collection centers and mobile testing vehicles in all six city districts (Fig. 6.3). Other efforts to improve testing capacity and access included workplace screenings, which were conducted in partnership with various businesses. A satellite monitoring approach, one that relied on an artificial intelligence system, was also used to follow-up on confirmed COVID-19 cases, monitor the progression of symptoms, and facilitate their referral to health facilities, as needed.

These efforts to bolster diagnostic testing were complemented by those geared toward developing contact tracing capacity. With the assistance of the Department of the Interior and Local Government and various barangay health units, more than 3500 individuals were trained as contact tracers. These individuals were mobilized to conduct interviews and initial public health risk assessments of COVID-19 cases and their close contacts, once test results had been received from various testing institutions. To support this effort, satellite contact tracing offices were built in each of Quezon City's six districts. Additional computers and vehicles were also procured to facilitate contact tracing activities and operations in the central and satellite command centers.

FIGURE 6.3 Quezon City Mobile Community Testing Units were deployed to COVID-19 hotspots throughout the city.

Contact tracing efforts also included unique public-private partnerships. One notable example included partnering with the private sector to develop a standardized, digital contract tracing mobile application, the Kyusi Pass Contact Tracing App, that facilitated compliance with occupancy limits, mask-wearing policies, and other considerations related to limited reopening. Following its development, and in partnership with the business sector, this app was implemented in all establishments in Quezon City. However, public adoption of the app was hindered by competing contact tracing app platforms, such as those implemented by neighboring local governments and the national government. Furthermore, siloed implementation and poor interoperability across these app platforms hampered the implementation of contact-tracing activities. The utility of these apps was also hampered by their reliance on mobile internet, which, as of 2021, is still beyond reach for an estimated 25.9% of Filipinos nationwide [10]. This unfortunate circumstance resulted in many business establishments resorting to maintaining handwritten logbooks for patrons and then inputting necessary contact tracing information.

To address several of the identified difficulties associated with contact tracing and to strengthen data collection and analysis, the QCHD's Epidemiological Surveillance Unit partnered with local research institutions—such as the University of the Philippines Diliman, the Ateneo de Manila University, and the OCTA Research Group—to analyze and monitor trends in disease transmission and the general health situation. Academics were invited to join the monthly city-led surveillance meetings, which enhanced the rigor of the response, and the quality of monthly reports that provided detailed metrics such as attack and reproduction rates.

Special task forces were also established to coordinate the surveillance and response operations of the LGU during outbreaks of variants of concern. For instance, upon the emergence of the Delta variant, the city established the Delta Variant Task Force. These task forces helped to identify areas within the city with the highest rates of disease transmission and facilitated the focused implementation of response activities, such as contact tracing, special concern lockdowns, and the provision of social services.

Pillar 3: isolation

Quezon City implemented both facility-based isolation and home quarantine interventions to respond to the COVID-19 pandemic. At the beginning of the pandemic, uncertainty surrounding the transmission of the virus resulted in an initial focus on facility-based isolation interventions. In the Philippines, Quezon City was one of the first LGUs to establish its own COVID-19 temporary treatment and monitoring facilities. Since March 2020, the city has established 12 of these facilities through cooperation with other government

FIGURE 6.4 QC HOPE Facilities were community-based isolation units operated directly by the city government. These were designed to care for confirmed COVID-19 presenting with mild symptoms or asymptomatically. In partnership with both business and public sectors, schools, hotels, and other suitable establishments were converted into fully equipped temporary treatment and monitoring facilities for COVID-19 patients.

agencies, educational institutions, humanitarian organizations, and business sector groups (Fig. 6.4). Additionally, 86 of the city's 142 barangays have constructed temporary structures or designated and repurposed existing buildings to serve as facilities for quarantine and isolation. This has resulted in 878 additional beds that improved the city's capacity to isolate and quarantine confirmed and suspected cases. Some of these facilities also offered specialized rooms and services, such as those designated for infected families and mother-child patient groups who required additional levels of privacy. Facilities for both quarantine and isolation were also set up for patients diagnosed with chronic renal disease and other immunocompromised patients, in collaboration with tertiary hospitals.

To complement facility-based isolation, guidelines for home quarantine were developed and implemented to cover recent domestic and international arrivals and people identified as close contacts of confirmed cases and suspected cases of COVID-19. Guidebooks were disseminated to inform households on the correct care and isolation of household members who were suspected COVID-19 cases. In addition, essential commodities such as food, infant necessities, household supplies, and hygiene products were provided to affected families, through subsidies from the local government and partnerships with nongovernmental organizations, sociocivic organizations, and from businesses (Fig. 6.5).

FIGURE 6.5   Reported cases of COVID-19, including those in hospitals, isolation units, and those under home quarantine, received COVID-19 care kits from the government and local partners. The kits displayed above were from the Philippine Red Cross-operated HOPE facilities, which contained hygiene supplies, vitamins and medicines, thermometers, pulse oximeters, and other basic necessities.

Pillar 4: treatment

The treatment-related actions implemented by the Quezon City LGU included implementing a triage and referral system, developing the health care workforce capacity to support treatment, and using telemedicine interventions. As previously referenced, Quezon City is home to several hospitals managed by the national government, the QCHD, and private sector actors. To ensure that these specialized care facilities were efficiently used, the city government coordinated with these hospitals through the National One Hospital Command, supervised by the National COVID-19 Inter-Agency Task Force. Through this effort, patients with severe disease were prioritized for admission to specialized care facilities in the city, while others with less severe disease were referred to lower-level city-owned facilities. Other facilities, such as the Red Cross Emergency Field Hospital, were also an important part of the COVID-19 referral network and served as a surge-capacity treatment facility for patients requiring only low-flow oxygen during outbreaks (Fig. 6.6).

Similar to the efforts in the Prevention Pillar, recruiting and retaining frontline health workers and training the Barangay Health Emergency Response Teams were critical in supporting Quezon City's treatment strategy. However, in addition to these efforts, the city also attempted to improve

FIGURE 6.6 The Red Cross Emergency Field Hospital was part of the COVID-19 referral network of Quezon City, serving as a surge-capacity treatment facility for patients requiring only low-flow oxygen during outbreaks.

treatment through telemedicine. Virtual medical consultations were conducted in partnership with businesses and volunteers, and their success generally depended on the availability of trained health care workers in the patient's area of residence, and the patient's access to the Internet. However, the city lacked the technological infrastructure and trained staff to institutionalize a system for teleconsultation. To address this challenge, as was the case with the development of the Kyusi Pass App, the city successfully collaborated with businesses to develop a technological solution. More specifically, it worked to develop an artificial intelligence system for digital health, BantAI COVID, that assisted in monitoring confirmed cases of COVID-19. The system was operated by a private firm staffed by volunteer doctors who were sourced by the Philippine Medical Association.

Quezon City also utilized its health promotion channels to promote health-seeking behaviors and raise awareness about the availability of treatments for COVID-19. To this end, efforts were made to design a comprehensive plan that was applied across all levels of government (i.e., from barangay to the city at large). This plan included strategies for referring patients to appropriate health facilities, providing assistance to financially vulnerable families, and running errands for basic necessities.

These efforts were generally successful but did not come without challenges. Particularly challenging was the provision of proper care and referring patients with severe cases of COVID-19. This problem was further exacerbated by periodic increases in case numbers due to emerging variants that resulted in rapid increases in cases. Meanwhile, in cases of treatment failure and eventual patient demise, the city worked to develop a protocol that could be used to manage both COVID-19 and non-COVID-19 deaths. These efforts involved collaborations between the QCHD, the Social Services Development Department, the Local Civil Registrar, and the National Department of Interior

FIGURE 6.7 An unclaimed suspected COVID-19 deceased person being brought to the Quezon City Crematorium. Quezon City is the only city in Metro Manila with its own facilities dedicated to the management of COVID-19 deceased patients.

and Local Government to identify how to handle the deceased safely and respectfully. In doing so, Quezon City became one of the first LGUs in the Philippines to develop a plan for storing and handling the remains of deceased persons in a safe and culturally respectful fashion (Fig. 6.7). Free cremation was provided, and the city sent refrigerated vans to selected hospitals and funeral homes for the temporary storage of the remains of the deceased when mortality surges exceeded the city's capacity to process the bodies in accordance with established sanitation protocols.

Pillar 5: reintegration

The strategic pillar focusing on reintegration encompassed multisectoral efforts within the city government to ensure social cohesion and recovery amidst the pandemic, which were beyond the usual scope of the QCHD. These efforts included shifting toward paperless and contactless transactions, creating a digital identification system for government services, providing assistance and social support to vulnerable families, and promoting business-friendly initiatives.

As part of a wider effort geared toward the digital transformation of all transactions, the city government implemented a "one-stop-shop" approach and fully automated the renewal of business licenses and other government transactions. Satellite district offices offering various government services were also established to decongest public transport by bringing public services closer to densely populated districts that were relatively distant from City Hall. Furthermore, all city offices were encouraged to use digital solutions to organize meetings, process documents for approval, and collaborate with each other, as a means of enabling staff who were remotely deployed, under quarantine, or unable to attend physically, to continue working with colleagues. These efforts helped sustain, or even increase, the overall productivity of the city government throughout the pandemic response.

The creation and implementation of a digital identification system, the "QCitizen ID," was also critical to the city's digital transformation (Fig. 6.8). This system was designed to ensure that all Quezon City residents had access to services offered by the city government and were accounted for in resource planning and allocation, including for those living in informal settlements. It was first used to identify households for social relief during the pandemic, especially in economically depressed neighborhoods and informal settlements. This was justified as, recognizing the substantial social and economic impacts of lockdowns and quarantines, the city's Social Services Development Department provided financial assistance to economically vulnerable households and also supported the families of individuals who had been sent to quarantine facilities. For example, children of quarantined individuals in the city were given access to programs such as play therapy. Further, as discussed later in this chapter, this system is envisioned as one that will underpin all future investments in big data management for health and social services in the city.

FIGURE 6.8   Honorable Mayor Joy Belmonte during the launch of the QCitizen Card, an all-access card allowing each citizen of the city to have digital access to all local government services.

The government-mandated lockdowns necessitated by the pandemic also threatened food security and the livelihoods of hundreds of families in Quezon City. This resulted in the city government reconfiguring undeveloped land and office rooftops into urban farms that provided fruit and leafy vegetables. Other efforts included establishing "container fisheries" that produced tilapia, catfish, and other local freshwater fish, with the assistance of the Department of Agriculture and the Bureau of Fisheries and Aquatic Resources. These farms and fisheries were then linked to end-user establishments and markets. Demonstration sites were set up within the City Hall complex to encourage and teach the public how to participate in urban farming and aquaculture. Other Filipino cities had similar initiatives—but not at the scale undertaken by Quezon City.

Aside from the financial assistance provided by the national and city governments to households affected by lockdowns, supplemental livelihood assistance and skills training sessions were also provided to existing small-to-medium enterprises, as well as those interested in starting a business. Quezon City's Small Business and Cooperative Office organized sessions on skills training, financial literacy, and business operations, while the City's Tourism Department held "Made in QC" trade fairs to showcase the wide array of products offered by new local business owners. These trade fairs have since made the rounds nationwide. Coffee table books and other promotional materials have also been produced to highlight the products of the city's small and medium business enterprises. At the time of writing, further commercialization of these products, including the use of e-commerce, is being explored.

An additional component of the city's reintegration strategy that was important was facilitating access to gainful employment among sectors that were heavily impacted by the pandemic. For example, in response to the various challenges associated with contact tracing efforts, as well as the closure or downsizing of various businesses, the city's Public Employment Service Office was tasked by the QCHD to recruit unemployed residents as contact tracers. In doing so, these efforts not only addressed the need to bolster contact tracing capacities but also assisted individuals who were in need of temporary employment because of the pandemic.

Looking ahead, by projecting an image of a safe city capable of handling public health emergencies, Quezon City has positioned itself as a viable location for businesses and investors, which could potentially lead to a faster economic recovery. The city government has prioritized regulations for the "new normal" that will follow the pandemic, as well as certifications that indicate a readiness to operate businesses while maintaining public health standards. The city government has also capitalized on its standing as an education and research hub and coordinated with local institutions to support these efforts and to develop evidence-based initiatives that have enabled the resumption of business operations for surrounding establishments and other dependent industries. More broadly, through its reintegration strategy, the city continues to facilitate employment among its working-age population, create a welcoming environment for international investment, and foster the growth of local enterprise. It is hoped that these efforts will help the city recoup economic losses incurred during the pandemic while simultaneously opening new avenues for economic growth.

Pillar 6: vaccination

The implementation of the national COVID-19 vaccination program was and continues to be a responsibility shared between the national government and local governments. While national-level planning and procurement ensured the equitable distribution of vaccines nationwide, cities, municipalities, and provinces were given the authority to implement policies and guidelines that were more nuanced and contextualized. Subnational levels of government were also given the option to procure additional vaccines, as needed.

Quezon City was uniquely situated to inform and steer both national and local vaccination efforts. The city was one of several pilot sites for the implementation of COVID-19 vaccination campaigns in the Philippines, which commenced in March 2021. Meanwhile, in compliance with national guidelines, the city's Vaccination Task Force—dubbed the "Vax to Normal" Team—developed its own COVID-19 Vaccine Plan called "QC Protektodo," a portmanteau of words that mean "QC Protects All." Additionally, a Vaccination Program spearheaded and housed within the QCHD and collaborated with local businesses, nonprofits, and academics to focus on a variety of efforts including policy development, capacity building, and data management for the issuance of vaccination certificates, among other tasks. As was the case with other areas within the Philippines, Quezon City's vaccination policies prioritized vulnerable population sectors, such as medical and nonmedical frontline personnel, people with comorbidities, and elderly persons. In total, 140 vaccination sites were established throughout the city in areas identified by the barangays, including public areas such as malls and train stations. These vaccination efforts were promoted and publicized using both mass and social media, and through other health promotion channels that had been made operational during the early months of the pandemic (Fig. 6.9).

Other innovative approaches championed by the city government included concerted efforts to provide vaccinations to specific demographic groups. For instance, *"Bakuna Nights,"* were concerted events that provided vaccinations after office hours and catered primarily to the city's working-age population. Other similar efforts were involved providing at-home vaccination services for bedridden individuals, high-risk elderly persons, and their families; as well as vaccinations in closed institutions such as jails, juvenile detention centers, religious convents, homes for the elderly, and those providing services for special populations such as the differently abled.

FIGURE 6.9 Social media was utilized enormously by the city to promote its vaccination efforts, under the QC Protektodo campaign.

Quezon City also worked to partner with nongovernmental institutions and actors to improve vaccination campaigns. For instance, partnerships with nongovernmental organizations, academic institutions, the private sector, and faith-based groups, led to the establishment of strategically located vaccination sites, the recruitment of human resources, and the ability to meet other logistical needs. Other efforts included a volunteer recruitment program that accepted professionals, students, and others from health and nonhealth fields to participate in the city's vaccine roll-out.

Of these, the partnerships that established strategically located vaccination sites proved especially useful for promoting vaccine uptake. International SOS—an independent private global health and medical service company—was engaged by international donors to undertake a review of the city's COVID-19 response. Unsurprisingly, the review found that rates of both diagnostic testing and immunization were highest in the areas most proximal to the testing and immunization facilities and that the immunization clinics with the best attendance were those situated in high-traffic areas such as shopping malls, markets, and churches.

As of June 30, 2022, the Quezon City's Vaccine Task Force reports that 98% of its target population has been fully vaccinated, with approximately half of the population also receiving one booster vaccination. Notably, this figure includes vaccinations administered in all health facilities based in Quezon City, including those that are not managed by the city government, such as hospitals managed by the national government and vaccination sites operated by private businesses. Collecting this data from institutions not under the city's jurisdiction has been a notable achievement, as access to vaccination-related data was required to comply with the Philippine Data

Privacy Act of 2012, and entailed extensive coordination with the National Department of Health and the Department of National Defense—the two national departments operating hospitals within Quezon City—that maintained different guidelines for the release of vaccination-related data.

Utilizing COVID-19 investments to build a stronger urban health system in Quezon City, Philippines

The capacity to prevent or mitigate outbreaks of COVID-19 and other emerging diseases is critical to ensuring the continued growth and prosperity of the Philippines' most populous city. In pursuit of a more resilient and sustainable postpandemic future, Quezon City partnered with International SOS, with the financial support of the International Finance Corporation and the Australian Government, to conduct a rapid assessment of the city government's COVID-19 response and to identify priority areas for investment.[2]

Conducted in March 2022, the rapid assessment gathered representatives from the Mayor's Office and other city government offices and departments in charge of investment, social services, budget, and management, as well as the health and public affairs and information services departments. This assessment aimed to help develop interventions that will enhance the government's capacity to protect its citizens from future epidemics and pandemics while ensuring financial sustainability and achieving broader public health goals beyond COVID-19. An assessment framework, grounded in the PDITR + V Framework and pillars, was used to analyze the city's level of achievement across 107 tasks recommended by the plan. Results showed that Quezon City was able to complete 68 of these recommended tasks and that an additional 14 were on track to be completed, while one item was beyond the scope of the city government.[3] Meanwhile, 24 tasks required further evidence for validation, which were eventually included in the report's recommendations, and informed the prioritization of areas for investment.

Remarkably, the city government was able to address the negative impacts of the pandemic on the city's health situation, economy, and livelihood despite various constraints. Among these were logistical challenges posed by the recommended tasks and an ongoing need to balance the implementation of national guidelines within the realities of local contexts.

To briefly summarize, the identified strengths of Quezon City's response included the prompt detection of possible cases of COVID-19, including variants of concern, through increased laboratory capacity and targeted mass

---

2. International SOS is a health and securities services firm that supports clients to maintain a resilient workforce so they can achieve business continuity and productivity, working from over 1,000 locations in 90 countries, with 13,000 staff.
3. More specifically, the item beyond the scope of the city government was ensuring health insurance coverage, which is the role of the country's national health insurer.

testing of vulnerable groups; the generation of data on testing, contact tracing, surveillance, and vaccine uptake to inform the city's evidence-informed pandemic response; the development of guidelines for quarantine, isolation, and treatment, which were sensitive to local needs; and the pursuit of partnerships between the public and business sectors through legislation, projects, and campaigns.

Ultimately, several opportunities for health system strengthening were also identified, with an emphasis on sustainability, integration, and multisectoral involvement. Specific interventions were then proposed by International SOS and guided by published examples of resource-efficient projects, activities, and campaigns. These interventions were then ranked by key stakeholders in the QCHD in terms of urgency and importance, with subsequent validation from senior representatives of the city government. Projects that were given the highest priority fell under three thematic groups—digital solutions and big data management, strengthening surveillance systems, and creating a dedicated health education and promotion unit—that are now described in greater detail.

Digital solutions and big data management for health

The COVID-19 pandemic highlighted the need for robust data science and systematic data management, especially in a city with a population of almost 3 million people. The successful use of data science in pandemic-related interventions and policy development was typified using publicly available data on human mobility, together with data on contact tracing, molecular testing, and health care utilization [11]. Meanwhile, with the impending national roll-out of Universal Health Care in the Philippines, Quezon City was at a critical juncture. It will need to implement a seamless digital platform for all health operations that not only will support its postpandemic surveillance and decision-making but also will achieve the objectives of the 2019 Universal Health Care Law.

Currently, despite the strengths of the city's innovative response, the city government's capacity to handle data for its large population is hampered by the lack of sufficiently trained personnel, equipment, and technological infrastructure. At the time of the rapid assessment, the city also lacked a unified digital ecosystem for data storage, processing, and visualization of health and economic indicators. This led to some delays in decision-making and upscaling of activities during the COVID-19 response in areas with greater need.

Proposed interventions in this thematic group that could be used to address these issues in the future include maintaining an online platform to book testing, medical consultations (i.e., physical or telemedicine), and vaccination services offered by the local government; creating a test result management system that allows for faster processing and release of results to the patients, contact tracing teams, quarantine/isolation facilities, and other health facilities; developing artificial intelligence to assist in medical triaging, isolation/

quarantine monitoring, community-based surveillance, and contact tracing; establishing a comprehensive system for interfacility referral, including digital document management, that operates across various levels of health care management; developing an electronic medical records system with harmonized implementation across all health facilities (i.e., private and public health facilities) within Quezon City; improving the compatibility of data used by the reimbursement systems of various health financing institutions and health maintenance organizations; pursuing data-driven quality improvements for health facility operations, supply chain mechanisms, inventories, and relevant budgeting and procurement processes, using insights from electronic medical records and health operations data; and creating an online dashboard containing real-time data (e.g., health statistics, risk assessments, and hazard maps) that could facilitate evidenced-informed decision-making and programs.

A key feature to be included in many of these proposed interventions is integration with the QCitizen ID, the city's innovative digital identification system. In its original plans for the QCitizen ID, the city government hoped that the ID would facilitate the consolidation of health and social services, such as electronic medical records, criminal records, insurance systems, social amelioration programs, special services for senior citizens, persons with disabilities, and other vulnerable populations. Building on investments in digital transformation that were initiated as part of the COVID-19 response, the city government further integrated its services with the QCitizen ID system, as a means of speeding up transaction times. These innovations helped build public trust and deter graft and corruption, while also lowering the risk of COVID-19 transmission among staff and clients. The digital transformation of government processes, such as meetings and document approvals, was also strengthened, which led to improvements in the overall productivity of the local government. More broadly, it is anticipated that the proposed interventions will further increase Quezon City's competitiveness as an investment destination in the postpandemic era.

Strengthening of surveillance systems

Robust surveillance systems are critical to the success of COVID-19 prevention and treatment efforts. During the pandemic, surveillance data guided the implementation and adjustment of COVID-19 control measures, including contact tracing, the isolation of cases, and the quarantine of close contacts and suspected cases. It also allowed the government to assess community quarantine protocols, which eventually enabled the safe resumption of economic and social activities. More broadly, surveillance data allowed key decision-makers to evaluate the overall impact of the pandemic on health systems and society.

The rapid assessment conducted by International SOS revealed the need for more dynamic surveillance systems for future threats and the assurance of overall health security in Quezon City. The results also highlighted the city's potential in averting future outbreaks in the country as it has already made significant investments in disease detection. Importantly, because of its geographic location (i.e., as a gateway to the country's northern provinces), and its status as a transportation, commerce, and education hub, the city finds itself in a good position to lead the charge for local-level disease prevention.

In the longer term, the city's surveillance team will continue to monitor the incidence of COVID-19 transmission, morbidity, and mortality within the city's population, with an emphasis on vulnerable and disadvantaged groups. At the time of writing, COVID-19 cases are on a gradual decline; nonetheless, the established surveillance systems are well-positioned to track potential epidemiological changes over time and to detect outbreaks of existing and emerging SARS-CoV-2 variants.

The results of the consultative process with the city government further reinforced the city's commitment to strengthen its surveillance efforts. In summary, the city has decided to prioritize strengthening existing surveillance systems, decentralizing epidemiologic surveillance unit functions, coordinating data management and analysis, and strengthening laboratory capacities. These efforts include continuing to report influenza-like illness and severe acute respiratory infection while increasing the utilization of sentinel reporting sites through facility enhancement, capacity building, and training. Existing systems will also be strengthened by mobilizing the public health workforce to conduct case finding and contact tracing, improve primary care surveillance units (i.e., barangay health centers) and sentinel surveillance sites, and implement regular training sessions. These training sessions will emphasize evolving case definitions, improving COVID-19 management at the community level, and strengthening the Quezon City health service network to improve the delivery of health services and response activities.

Additional efforts to decentralize epidemiologic surveillance unit functions will focus on the district level. This grassroots-focused approach for conducting surveillance and analysis is expected to speed up data analysis and allow for more localized public health interventions in response to health events and notifiable diseases. As such, more resources are starting to be allocated to strengthen district satellite surveillance units, allowing them to perform sample collection, testing, and reporting. Resources are also being provided to support these units as they train community-level disease surveillance officers and volunteers. The epidemiological surveillance unit continues to be the ultimate data collection point for various health parameters, providing support to executive and legislative offices for different city-wide health programs. Further, the city will seek to improve the coordination of data management and analysis from all health facilities in Quezon City, including those in the private health sector.

Laboratory and testing capacities will also be strengthened in the city through several efforts, including the creation of a transition plan for the COVID-19 molecular laboratories that will seek to enhance and repurpose them to allow for the detection of other emerging diseases. Quezon City and the rest of the country face the relentless threat of other emerging and re-emerging infections, such as monkeypox and influenza; infectious diseases of public health significance, such as dengue, measles, HIV/AIDS, hepatitis, tuberculosis, and gastrointestinal pathogens; as well as vaccine-preventable diseases, such the human papillomavirus that causes cervical and nasopharyngeal cancers. Because of this, the city government has prioritized strengthening laboratory capacities that promote the detection of other diseases. For example, the city's molecular laboratory is currently repurposing its cartridge-based nucleic acid diagnostic systems to detect the aforementioned pathogens. Other proposals for the detection of antimicrobial resistance, environmental disease surveillance, and precision medicine testing are also being considered for future implementation, pending regulatory and policy support from the Department of Health.

Quezon City Health Education and Promotion Unit

In 2019 the Philippine government passed the Universal Health Care Act (Republic Act 11223), which aimed to enhance the capacity of the health system to provide quality health care while ensuring financial risk protection. Since its passing, the Act has led to the development of a strategic reform agenda that directs the national government and local governments to strengthen health promotion and communication surrounding health risks and hazards.

Despite being the country's most populous city, Quezon City did not have its own Health Education and Promotion Unit before the COVID-19 pandemic, and functions related to health promotion were instead nested under the QCHD. The pandemic eventually caused the QCHD to focus on client-facing activities related to the pandemic response, and all efforts related to health promotion and risk communication were delegated to different offices, most notably the city's Public Affairs and Information Services Department. While the two city departments coordinated on key messaging and strategy, the lack of technical oversight from the QCHD had an undeniable impact on the city's health promotion and risk communication activities. A new Quezon City Health Education and Promotion Unit within the QCHD is, therefore, being developed to address these challenges.

The Health Education and Promotion Unit will be charged with several key objectives including developing directions, policies, standards, and guidelines pertaining to health promotion; providing policy advice to partner agencies related to the key health determinants; providing technical assistance to the city administrator, the city council, other LGU departments, partner national agencies, and health facilities and institutions on health promotion and risk communication; leading the implementation of national campaigns and health messaging initiatives at the local level, while accounting for the specific contexts of the city; providing health promotion data to different stakeholders; producing and providing health information, education, and communication materials for different projects, activities, and campaigns; and establishing networks with local partners and community-based organizations, especially those involved in peace, shelter, livelihood, education, food, income, ecosystem, social relations, equity, poverty, social justice, the empowerment of women, and human rights.

At the time of writing, the QCHD has submitted a proposed structure of the unit to the city council, which, if approved, would establish the office, create permanent positions, and deliver the mandate of the Health Education and Promotion Unit. While awaiting approval from different national agencies for budgetary and technical support, the QCHD has already designated staff from other existing programs to perform the functions of the unit. Ultimately, by pursuing a dedicated unit for health promotion, the city is progressing toward the implementation of activities based on the actual needs of the city, increased accountability in the implementation of different projects and campaigns, and a more synergistic, inclusive, and coordinated approach in health and social services.

Conclusion

Quezon City's pandemic response may be summarized in two words: innovative and collaborative. Despite unprecedented challenges, the city successfully implemented multisectoral strategies toward prevention, detection, isolation, treatment, and reintegration in ways that not only fulfilled national guidelines but also addressed local needs.

The success of Quezon City's pandemic response highlighted the importance of multisectoral collaboration, stakeholder engagement, and the strengthening of institutional and investment linkages. The city government was able to successfully stimulate discussion between investors, business owners, heads of offices and organizations, sectoral representatives, and academic and research institutions, thereby enabling collaborative and innovative solutions—such as the establishment of vaccination sites in nontraditional locations and the recruitment of volunteers from nontraditional backgrounds. There is little reason to believe that these successes cannot be enjoyed by other local governments, as long as they are well-connected with stakeholders and partners and have a keen awareness of the needs and vulnerabilities of the local population, which are informed through robust data and meaningful engagement.

The rapid assessment of Quezon City's COVID-19 response revealed several strengths in the city's response including the prompt detection of possible cases of COVID-19 through increased laboratory capacity and targeted mass testing of vulnerable groups, the development of locally relevant guidelines for quarantine, isolation, and treatment, and partnerships between the public and business sectors. Nonetheless, the experiences from the pandemic laid bare vulnerabilities and areas for further health systems strengthening and investments, such as the need for big data management and analysis, improvements to surveillance systems, and a more focused and coordinated approach to health promotion. It is remarkable that, only a few months after the rapid assessment was completed, Quezon City's leadership took concrete steps to address these gaps and is now on track to accomplishing these valuable investments.

Further, many of the innovations that were introduced during the COVID-19 pandemic involved human and capital investments. These investments should enable the city to "future-proof" itself against future epidemics and pandemics, and be at par with other cities in industrialized countries in ensuring preparedness. More importantly, the experiences throughout the pandemic response have inspired a sense of hope and confidence in the city's leadership. In the words of Mayor Belmonte at the end of 2021, after 2 years of leading a successful initial response against the outbreak, "any pandemic state is volatile, but at least confidence inspired by preparedness has replaced fear and anxiety." It is this confidence that will further lead to increasing returns, as Quezon City continues to enhance its economic competitiveness, health resilience, and social capital, as it aims to be the Philippines' foremost investment gateway following the COVID-19 pandemic.

References

[1] Quezon City Government. Annual Report (June 2020–June 2021): Navigating the road to recovery. Quezon City: Quezon City Government; 2021.

[2] Pante MD. Quezon's City: corruption and contradiction in Manila's prewar suburbia, 1935–1941. Journal of Southeast Asian Studies 2017;48(1):91–112. Available from: https://doi.org/10.1017/S0022463416000497.

[3] Zero Extreme Poverty PH 2030. Quezon City, <https://zeropovertyph.net/quezon-city_profile/>; 2018 [accessed 16.10.22].

[4] Mateo J. Belmonte, Sotto win second terms in Quezon City. The Philippine Star 2022;11.

[5] Dioquino AH. Defensor challenges Belmonte to a debate. Manila Bulletin 2022;2.

[6] Simeon LM. African swine fever confirmed in third Quezon City barangay. The Philippine Star 2019.

[7] Moaje M. Covid-19, dengue cases on the rise in Quezon City. The Philippine Times 2022;08.

[8] Chu DT, Vu Ngoc SM, Vu Thi H, Nguyen Thi YV, Ho TT, Hoang VT, et al. COVID-19 in Southeast Asia: current status and perspectives. Bioengineered. 2022;13(2):3797–809. Available from: https://doi.org/10.1080/21655979.2022.2031417.

[9]   Quezon City Government. QC COVID-19 update − as of June 30, 2022−8AM, <https:// quezoncity.gov.ph/covid19counts/qc-covid-19-update-as-of-june-30-2022-8am/>;   2022 [accessed 16.10.22].

[10]   Statista Research Department. Mobile phone internet user penetration in the Philippines from 2017 to 2025, <https://www.statista.com/forecasts/975001/philippines-mobile-phone-internet-user-penetration>; 2022 [accessed 16.10.22].

[11]   Shah N, Steinhardt J. How data science can ease the COVID-19 pandemic. Washington, DC: Brookings; 2020.

Section III

Workforce and surge capacity

What is surge capacity?

The Agency for Healthcare Research and Quality defines surge capacity as, "a health care system's ability to rapidly expand beyond normal services to meet the increased demand for qualified personnel, medical care, and public health in the event of bioterrorism or other large-scale public health emergencies or disasters" [1]. Building surge capacity requires robust workforces, equipment supplies, and facilities, as well as coordination across health systems [2,3]. Health care facilities must be able to handle high patient volumes, public health systems must be able to mobilize risk communication, contact tracing, and vaccination campaigns, and communities must be prepared to comply with non-pharmaceutical interventions [1]. Widespread failures to the increased demand on health care systems around the world have underscored the importance of surge capacity throughout the response to COVID-19 and for building resilient health care systems and communities in the future.

What constitutes strong surge capacity?

Good governance, as outlined in Section I, is essential to surge capacity. Scaling-up public health and clinical capacity in a matter of days requires coordination at all levels of government, public health systems, and hospital staff. For instance, one of the following chapters describes how the Southern Nevada Health District conducted tabletop outbreak simulations that improved emergency operation plan-based coordination between health care and government organizations during COVID-19 in Las Vegas.

Rapidly expanding clinical and public health workforces is also essential to surge capacity. In clinical settings, health care workers must respond to drastic increases in patient volume and health care worker absenteeism due to exposure/illness, while continuing to manage regular emergent cases [2]. Public health staff must rapidly mobilize risk communication campaigns with community members, initiate contact tracing efforts, and conduct analyses to make appropriate policy recommendations. Jurisdictions may use creative strategies to find additional staff. During the COVID-19 pandemic, retired health care professionals, early medical graduates, people with foreign medical licenses, and military health care workers were brought in to manage surging patient needs [4]. To expand contact tracing teams, university students, individuals facing unemployment due to pandemic-related closures, and members of the military were trained and hired. Leaders in Dhaka expanded their electronic payment program to ensure that community health workers could be paid for their additional hours, and transfer hours normally spent traveling to the bank on patient visits.

The ability to rapidly expand the number of hospital beds available was also crucial when inpatient care volumes rose rapidly. Doing so required rescheduling of nonemergent appointments, expanded use of telemedicine, mobile and temporary hospitals, and referral protocols [2]. These efforts were paired with resource coordination, as increased bed capacity was not helpful if not accompanied by sufficient medication, oxygen, and other resources.

Why does surge capacity matter?

Failures during the COVID-19 pandemic illustrated the importance of building surge capacity. Throughout the outbreak, clinical staffing, resources, and hospital facilities were overwhelmed by infections among workers and increased patient volume, particularly in critical care areas. Severe shortages of personal protective equipment, ventilators, and hospital beds resulted in excess mortality during the first months of the pandemic [5]. Hospital shortages forced the cancellation of regular medical appointments, screenings, routine vaccinations, and elective surgeries, further increasing morbidity and mortality associated with limited surge capacity [6,7]. Patients delayed seeking care for emergent conditions including stroke, heart attacks, and hypoglycemia [8]. In many countries, contact tracing programs floundered under high case totals and limited public cooperation [9].

The importance of surge capacity urban environments

Urban centers played an outsized role in supplying surge capacity. Highly transmissible diseases spread faster and exceeded existing clinical and public health capacities in places with high population densities. Furthermore, large hospitals with specialized care centers were often located in cities, supporting those in nearby peri-urban and rural areas in addition to the immediate community [10]. Individuals from nearby regions traveled to urban centers to seek specialized care because city hospitals were better resourced [11]. Clinical and public health systems needed to be prepared to respond to these additional stressors. Regional leaders had tools to address these additional challenges, such as increasing the resources available to nonurban health facilities, building additional infrastructure with beds for overflow patients, and waiving telemedicine regulations [1,12,13].

Chapter introductions

The chapters in this section focus on the efforts to surge the health care workforce in Las Vegas, Nevada, United States of America and Dhaka, Bangladesh as a means of supporting the response to the pandemic. The health systems in both cities faced overwhelming demand during COVID-19, and as a result, the capacities were surged by adapting management. These examples illustrate how a combination of traditional and nonconventional approaches tailored to their communities can drastically expand the reach of health care workers and public health staff.

The chapter on Las Vegas addresses the approach used for leveraging public and private partnerships between local, state, tribal, and federal agencies to respond to the COVID-19 pandemic (Fig. 1). These team-oriented efforts led to enhanced communications, improved operational coordination, and trust among those involved in the pandemic response, and better access by the public to much-needed health resources and services. This surge in

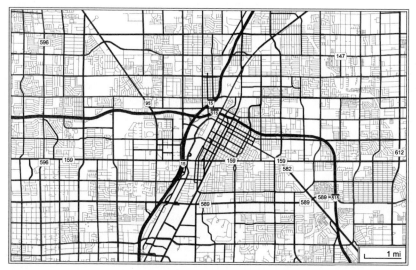

FIGURE 1 A map of Las Vegas, Nevada, United States of America. Base map sourced from OpenStreetMap, using materials made available under CC BY-SA 2.0.

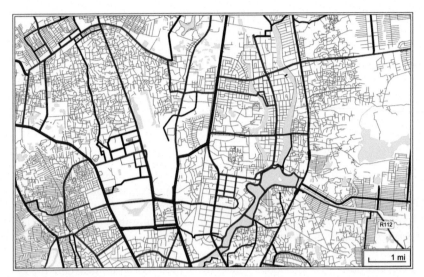

FIGURE 2 A map of Dhaka, Bangladesh. Base map sourced from OpenStreetMap, using materials made available under CC BY-SA 2.0.

capacity required planning and active participation by diverse stakeholders who shared a mission to protect the public's health and to support ongoing recovery in the city and surrounding areas.

In Dhaka, Bangladesh, the COVID-19 pandemic exposed weaknesses in the city's infrastructure, leading to a high number of cases and death counts (Fig. 2). To blunt these negative impacts of the pandemic on the urban poor, BRAC, one of the largest NGOs in the world, used a previously established cadre of community health workers to surge the workforce as a means of delivering crucial health care services, raising awareness, and gathering data. Importantly, these efforts were aided through the use of mobile financial services, which were used to facilitate data gathering and ensure effective operations amidst ongoing lockdown and social isolation measures.

References

[1] Koh HK, Shei AC, Bataringaya J, Burstein J, Biddinger PD, Crowther MS, et al. Building community-based surge capacity through a public health and academic collaboration: the role of community health centers. Public Health Reports 2006;121(2):211−16. Available from: https://doi.org/10.1177/003335490612100219.

[2] Kumar P, Kattan O, Broome B, Singhal S. Reassessing Covid-19 needs: how providers can reexamine their surge capacity, supply availability, workforce readiness, and financial resiliency. New England Journal of Medicine Catalyst Innovations in Care Delivery 2020;. Available from: https://doi.org/10.1056/CAT.20.0112.

[3]   Adams L.M. Exploring the concept of surge capacity. Online Journal of Issues in Nursing. https://doi.org/10.3912/OJIN.Vol14No02PPT03.

[4]   Williams GA, Maier CB, Scarpetti G, Giulio de Belvis A, Fattore G, Morsella A, et al. What strategies are countries using to expand health workforce surge capacity during the COVID-19 pandemic? Eurohealth 2020;26(2).

[5]   Rubinson L, Shah C, Rubenfeld G. Surge mechanical ventilation for the COVID-19 surge and future pandemics—time to reframe the strategy. JAMA Network Open 2022;5(8): e2224857. Available from: https://doi.org/10.1001/jamanetworkopen.2022.24857.

[6]   Boutros M, Moujaess E, Kourie HR. Cancer management during the COVID-19 pandemic: choosing between the devil and the deep blue sea. Critical Reviews in Oncology/Hematology 2021;167:103273. Available from: https://doi.org/10.1016/j.critrevonc.2021.103273.

[7]   Dinleyici EC, Borrow R, Safadi MAP, van Damme P, Munoz FM. Vaccines and routine immunization strategies during the COVID-19 pandemic. Human Vaccines & Immunotherapeutics 2021;17(2):400−7. Available from: https://doi.org/10.1080/21645515.2020.1804776.

[8]   Lange SJ, Ritchey MD, Goodman AB, Dias T, Twentyman E, Fuld J, et al. Potential indirect effects of the COVID-19 pandemic on use of emergency departments for acute life-threatening conditions—United States, January−May 2020. MMWR 2020;69(25):795−800. Available from: https://doi.org/10.15585/mmwr.mm6925e2.

[9]   Lewis D. Why many countries failed at COVID contact-tracing—but some got it right. Nature 2020;588(7838):384−7. Available from: https://doi.org/10.1038/d41586-020-03518-4.

[10]   World Health Organization. Framework for strengthening health emergency preparedness in cities and urban settings. Geneva: WHO; 2021.

[11]   Chakrabarti S, Tatavarthy AD. The geography of medical travel in India: differences across states, and the urban-rural divide. Applied Geography 2019;107:12−25. Available from: https://doi.org/10.1016/j.apgeog.2019.04.003.

[12]   Monaghesh E, Hajizadeh A. The role of telehealth during COVID-19 outbreak: a systematic review based on current evidence. BMC Public Health 2020;20:1193. Available from: https://doi.org/10.1186/s12889-020-09301-4.

[13]   Forum on Medical and Public Health Preparedness for Catastrophic Events, Board on Health Sciences Policy, Institute of Medicine. Regional disaster response coordination to support health outcomes: summary of a workshop series. Washington, DC: National Academies Press (US); 2015.

Chapter 7

Expanding workforce surge capacity and the multijurisdictional response to the COVID-19 pandemic in the Las Vegas metropolitan area

Fermin Leguen, Cassius Lockett and Jeffrey Quinn
Southern Nevada Health District, Las Vegas, NV, United States

Background

Las Vegas metropolitan area (Clark County, Nevada)

With an estimated population of 2,350,206 persons and a 20.5% population increase from 2010 to 2022, Clark County, Nevada is one of the United States' largest and fastest-growing counties, comprising 73% of the state's residents and spanning over 7800 square miles [1]. Clark County is a diverse community with a unique set of challenges given its size and complexity.

Home to Nevada's largest city, Las Vegas, 54% of the county's population is White Non-Hispanic, while 32.8% are of Hispanic origin, 12.8% Black Non-Hispanic, 10.2% Asian, 6% multi-race, 1% American Indian/Alaskan Native, 1% Hawaiian/Pacific Islander, and 15% other [2]. In 2021 the median household income in Clark County ($70,075) was lower and the unemployment rate among the civilian labor force (5%) was higher when compared to the national averages [3]. The same US Census data suggest that there are some 302,606 individuals, or approximately 13% of the county's population, living in poverty, and 22.3% of adults are without health insurance.

Gaming, accommodation, and food services employ the largest numbers of residents in Clark County, followed by government and retail trade [4]. While employers in food services and retail trade are less likely to offer health insurance to employees in general, the employer-sponsored Culinary Health Fund offers full coverage health benefits with no deductibles or family-coverage premiums to over 130,000 Culinary Union members and dependents in the Las Vegas area. Other key industries in Southern Nevada include logistics and manufacturing, reflecting Nevada's business-friendly regulatory climate, anchor sectors (e.g., tourism and gaming), and proximity to West Coast population centers and transportation routes, However, the health care sector has remained relatively stable in Clark County, with limited access to care. An economic downturn that began in 2008 and the ensuing economic recession had a negative impact on gambling spending resulting in significant damage to the casino industry in America—ultimately undermining its future growth and potential [5]. Additionally, due to limited local and state funding investments, workforce education in Southern Nevada lags national benchmarks.

The city of Las Vegas is often impacted by population surges—sometimes as large as 200–300 thousand visitors—on weekends or for large events. In 2019 prior to COVID-19 Pandemic, Las Vegas received 42 million visitors over the course of the year. Annual visits were reduced to 19 million in 2020 during the pandemic but rebounded to 32 million visitors by the end of 2021, representing a ratio of about 14 visitors per local resident [6]. Due to risks associated with a large volume of visitors and COVID-19 transmission the city government, country government, and the Southern Nevada Health District (SNHD) developed a plan to isolate travelers that tested positive for COVID-19 but do not meet hospital admission criteria during their stay in Las Vegas. As a part of this plan, COVID-19-positive travelers were not allowed to board airplanes to return home until it was determined that they were no longer contagious. Some hotels discharged COVID-19-positive patients from their properties due to perceived risks to other guests. This issue was addressed through a COVID-19 Visitor Isolation Plan that contracted off Las Vegas Strip hotel locations willing to house visitors who needed to complete the COVID-19 recommended isolation period [7]. Costs for these services were supported through emergency federal legislation and funding provided through the CARES Act.

The public health system in Las Vegas, Nevada

Established in 1962, the SNHD serves as the local public health authority for the Southern Nevada region which includes all of Clark County and the incorporated cities of Las Vegas, North Las Vegas, Henderson, Mesquite, Boulder City, and Laughlin. The mission of the SNHD is "to assess, protect and promote the health, the environment, and well-being of Clark County residents and visitors." The SNHD safeguards the health of the community's residents and visitors through innovative public health programs, regulations, and initiatives focused on protecting and promoting their health and well-being [8]. The Southern Nevada District Board of Health (the "Board") is the governing body of the SNHD within Clark County, Nevada. As Health

District's governing body, the Board is vested with jurisdiction over all public health matters within Clark County.

Health care resources in Southern Nevada

Southern Nevada is served by 21 acute care hospitals, seven mental health hospitals, and 40 skilled nursing facilities [9]. Six public fire departments provide emergency medical services (EMS) in the region, which include the Boulder City Fire Department, Clark County Fire Department, Henderson Fire Department, Las Vegas Fire & Rescue, Mesquite Fire & Rescue, and North Las Vegas Fire Department. In addition, there are six private franchised EMS agencies serving the area, including two air ambulance services.

There are a total of 97,560 individuals working in health care and social assistance across Clark County [4], including 5,056 licensed physicians, amounting to roughly 218 licensed physicians per 100,000 population [10]; additionally, there are a total of 1,567 licensed Advanced Practitioners of Nursing and 683 Physician Assistants [10]. These numbers rank Nevada as one of the worst (45 out of 50) in the United States for active physicians per 100,000 population. The clinical workforce is even more strained when considering the additional population of tourists. Increasing the number of active physicians will have a positive impact on access to care and enhance residents' ability to receive care, including preventive services such as screenings and routine check-ups. To help address this need, in October 2022, the University of Nevada, Las Vegas (UNLV) opened the Kirk Kerkorian School of Medicine building on a nine acre campus that currently hosts over 200 medical students and over 300 residents and fellows.

Public health authority in Nevada

The power of administration of public health is defined in the Nevada Revised Statutes (NRS) 439, Administration of Public Health. More specifically, the SNHD's public health authorities are granted in NRS 439.361 to 439.3685, and NRS 439.366 stipulates the powers and jurisdictions of district boards of health (Box 7.1) [11].

Las Vegas remains both a national and international travel destination for millions of people each year. Accordingly, biological threats to public health remain high and the SNHD systematically monitors global emerging threats, informs community and health care partners, and prepares local responders and the health care workforce to respond to outbreaks and pandemics. Following the National Response Framework [12], and based on current, local, regional, tribal, and state emergency operation planning, written emergency operation plans (EOPs) are validated and used with local Homeland Security Threat Hazard Identification and Risk Assessments to better assist jurisdictions in prioritizing threats facing their resident and visitor populations [13]. As outlined in local EOPs, the SNHD is the lead response agency

---

BOX 7.1 Nevada Revised Statutes 439.366, which stipulate the specific powers and jurisdictions of district boards of health.

1. The district board of health has the powers, duties, and authority of a county board of health in the health district.

2. The district health department has jurisdiction over all public health matters in the health district.

3. The district health department:

(a) Shall, upon the request of the Nevada Gaming Control Board, advise and make recommendations to the Board on public health matters related to an establishment that possesses a nonrestricted gaming license as described in NRS 463.0177 or a restricted gaming license as described in NRS 463.0189 in the health district.

(b) May, upon the request of the Nevada Gaming Control Board, enforce regulations adopted by the Board concerning matters of public health against an establishment that possesses a nonrestricted gaming license as described in NRS 463.0177 or a restricted gaming license as described in NRS 463.0189 in the health district.

4. In addition to any other powers, duties, and authority conferred on a district board of health by this section, the district board of health may by affirmative vote of a majority of all the members of the board adopt regulations consistent with law, which must take effect immediately on their approval by the State Board of Health, to:

(a) Prevent and control nuisances;

(b) Regulate sanitation and sanitary practices in the interests of the public health;

(c) Provide for the sanitary protection of water and food supplies;

(d) Protect and promote the public health generally in the geographical area subject to the jurisdiction of the health district; and

(e) Improve the quality of health care services for members of minority groups and medically underserved populations.

---

for biological threat emergencies in the region. Other governmental agencies, health care, EMS, community partners, and private sector organizations are designated to specific tasks according to their respective emergency support and recovery functions, with shared responsibilities to support declared emergency or community response efforts. EOPs are evaluated or revised to incorporate lessons learned and corrective actions following after-action reports and components of EOPs are tested during community drills and exercises, several of which are discussed later in this chapter.

The Clark County government and the municipalities within the county are actively engaged in public health through the participation of designated elected officials from each jurisdiction in the District Board of Health. In addition, they have a strong influence on the impact of local public health interventions through local ordinances and through their control of financing, building codes, business licensing, and other considerations that are directly or indirectly linked to public health.

Lessons learned from previous infectious disease outbreaks in Southern Nevada

In the last 30 years, one of the first published large-scale infectious disease outbreaks in Las Vegas went undetected [14]. In 1993 a retrospective case-control study identified over 50 cases of bloody diarrhea in patrons that consumed hamburgers from at least eight fast food locations. Subsequent laboratory analysis of the hamburger revealed *Escherichia coli* O157:H7. Matched odds ratios of 9.0 found a strong association between the food and ill persons. This experience highlighted the need for laboratories in Clark County to have the capability to culture for reportable diseases like *E. coli*. Further, since this was a statewide multiagency effort that required epidemiological expertise from Oregon and the Centers for Disease Control and Prevention (CDC), an additional lesson learned was the need to create a local epidemiological surveillance unit and public health lab to address the public health needs of the community.

Since that time, the SNHD has responded to dozens of large-scale food-borne investigations in restaurants, picnics, weddings, and other social gatherings. One notable outbreak involved hundreds of ill persons that dined at a popular restaurant in 2013. Over 330 cases of foodborne illness caused by *Salmonella* bacteria were identified over a span of 6 weeks [15]. At least 50% of the ill persons sought medical care and approximately 15% were hospitalized. Using a case-control study design an analysis revealed that at least one food item was linked to illness. There was also evidence of cross-contamination. During this investigation, authorities at the SNHD gained experience in collaborating and learned how to work more efficiently and strategically with partners, including environmental health and public health laboratories. Another lesson learned through this experience was how to use epidemiology to guide testing and case interviews, especially in resource-limited contexts. More importantly, the SNHD gained valuable experience in using molecular methods, such as pulse-field gel electrophoresis, to differentiate between various groups and sources of possible *Salmonella* infection. This experience eventually helped the public health laboratory team further develop their molecular epidemiology skills—skills that were later used to implement whole genome sequencing methods that allowed for identifying specific COVID-19 variants circulating in the community.

In 2013 SNHD's epidemiology team responded to a complex cluster of tuberculosis (TB), with five laboratory-confirmed cases linked to a local hospital's neonatal intensive care unit. In collaboration with the CDC, the SNHD team investigated the cluster of TB cases identified at this facility.

The confirmation of these cases required a rigorous review of medical and TB clinical records. Unfortunately, in this instance, the SNHD was unable to obtain an isolate from the index case; doing so is important for communicable disease investigations, as it enables public health laboratories to genotype specimens and make genotypic comparisons between organisms and isolates from other cases and clusters. Still, one lesson learned from this investigation included the importance of linking local databases to other electronic data systems to improve the clinical management of TB cases and contacts. This skill became very useful during the pandemic because our surveillance team was able to enhance COVID-19 contact tracing investigations and identify cases and individuals at risk after linking multiple databases from local hospitals, commercial and public health labs, and the SNHD surveillance database.

More recently, between March and August 2017, an outbreak of Legionnaires' disease was investigated at a hotel facility in Las Vegas. As a part of this response, over 270 individuals were identified as being exposed to the *Legionella* bacteria; of those, seven were identified as confirmed cases, with an additional 31 individuals classified as suspect cases. This investigation highlighted the importance of educating hotel industry representatives about the importance of complying with preventative public health measures, including interventions in facilities with large-scale water systems and public hot tubs to implement water management plans and prevent the growth and spread of *Legionella* bacteria. Today, SNHD disease investigators and environmental health practitioners are often invited to national conferences to share best practices for investigating and controlling *Legionella*. In addition, in collaboration with the University of Nevada, the SNHD's Environmental Health Department hosts seminars on *Legionella* basics including disinfection, remediation, water management, and investigation.

The lessons learned from these outbreaks—whether related to establishing local capacities, using epidemiology to inform strategy in resource-limited settings, linking local databases to other data systems, using molecular techniques to identify and link the same organism in a food specimen and human, or educating industry representatives about the importance of public health measures—have ultimately helped to prepare Clark County and the city of Las Vegas for future public health events and emergencies.

Pandemic preparedness plans and simulation exercises

The SNHD and the Southern Nevada Health Care Preparedness Coalition developed emergency response plans using the Comprehensive Planning Guide (i.e., CPG 101) developed by the Federal Emergency Management Agency (FEMA) [16]. In addition, the SNHD has followed the Department of Homeland Security's Exercise and Evaluation Program guidelines for planning and conducting exercises [17].

Since 2015, there have been several notable exercises and training events to improve pandemic preparedness in Clark County. In September 2015, there was a full-scale exercise testing investigation and response to the intentional release of a biological threat, *Francisella tularensis* (Tularemia). Over 45 individuals from public federal, state, tribal, local agencies, resorts, and the business community participated in this exercise. The exercise tested several core capabilities including emergency operational coordination, public health and medical services, public information and warning, environmental response/health and safety, public and private services, and resource coordination.

The next year, in June 2016, a legal isolation and quarantine tabletop exercise was held. This discussion-based exercise reviewed current Isolation and Quarantine Bench Book reference documents with legal community and law enforcement agencies. The SNHD also provided an update on Ebola virus disease and other communicable diseases, a high-level overview of Isolation and Quarantine Plan processes in Nevada, which included a discussion of voluntary and involuntary isolation and quarantine, as well as the coordination of legal processes with district courts.

In February 2017, the SNHD held a seminar on closed points of dispensing in which risk management, security, safety, and human resource personnel from some of Las Vegas's largest public and private employers were invited to learn about public health emergency planning. This seminar discussed how medical countermeasures and other resources geared toward enhancing disease surveillance would be provided following a public health emergency, such as a pandemic.

In March 2020, the SNHD conducted a functional exercise on information sharing. Among other objectives, this exercise sought to test communication and information sharing, to submit operating status and capability assessment report forms to medical surge support teams, to engage and use emergency operations center software, to review incident command system (ICS) action and resource request forms, and to test radio, internet, landline, and satellite communications as alternate forms of interoperable and redundant communications.

Later, during the COVID-19 pandemic, as a part of Las Vegas COVID-19 mass vaccination campaign preparations, a drive-up influenza vaccine point of dispensing full-scale exercise was conducted in October 2020. This exercise tested several FEMA National Response Framework core capabilities including public health, health care, EMS/medical, medical countermeasure dispensing, and administration [12]. The exercise also tested the time required for the notification and assembly of response personnel, the setup and demobilization of points of dispensing, the throughput time of vehicles for medical countermeasure and dispensing operations, and site safety plans related to post-vaccination observation of populations in motor vehicles.

COVID-19 in Las Vegas, Nevada

During emergencies, the activation of the ICS can occur before the impacts of an emergency are felt locally. Such was the case with the COVID-19 pandemic, when the SNHD ICS was activated on February 10, 2020—3 weeks before the first confirmed COVID-19 case was reported in the United States. The SNHD also began to screen and monitor travelers returning from mainland China at McCarran International Airport (now Harry Reid International Airport), in February 2020. Known for being an innovative and technology-driven health department, the SNHD modified the existing Ebola symptom monitoring electronic application to enroll and monitor travelers in its local self-monitoring program. This application allowed individuals to report symptoms and temperatures twice daily for the duration of the monitoring period, which was determined based on guidance from the CDC. The CDC periodically reviewed and revised its recommendations and guidance related to the time required to isolate following a positive COVID-19 test. Early in the pandemic, isolation time recommendations were lengthier, but these recommendations were revised as more data became available.[1]

The first confirmed case of COVID-19 in Clark County was reported on March 5, 2020. On March 14, 2020, the Clark County Regional Policy Group—a group comprised of elected officials, first responders, industry representatives, and local government agencies such as the SNHD, Clark County School District, McCarran International Airport, and other partners—was activated to provide clear and fluid guidance. The group met daily during critical periods of the COVID-19 pandemic, allowing response organizations to share information that facilitated the effective activation of agency, local, tribal, state, and federal emergency operations plans, as well as the allocation of resources necessary to respond.

As soon became standard practice, members from the SNHD surveillance team followed each confirmed case throughout the duration of the isolation period recommended by the CDC [18]. Isolation of COVID-19 confirmed cases occurred either at home, hospitals, SNHD isolation facilities, City of Las Vegas isolation facilities or at dedicated hotel rooms contracted by Clark County. Once completed, each case was released from isolation and received isolation completion letters from the SNHD. Initially, cases were interviewed, and contact tracing was performed by SNHD staffing and contractors, as capacity allowed. As the outbreak escalated, however, case reports grew exponentially, which quickly exceeded the existing capacities necessary for properly managing all reported cases. Prior to the COVID-19 pandemic, the SNHD's disease surveillance team monitored approximately 20,000 electronic laboratory reports per month, but this number grew to over 300,000

---

1. Initially, the recommended isolation period was 14 days, later it was revised to 10 days, and, at the time of writing, the CDC recommendation is for an isolation period of 5 days [25].

electronic laboratory reports during surge periods of the COVID-19 pandemic. As a result, the SNHD's information technology systems were upgraded to handle this much larger volume of incoming reports. Other technology-based enhancements included modifying the reporting system for notifiable diseases to integrate data from multiple sources such as hospitals, commercial laboratories, medical offices, and other facilities. This modification allowed for the surveillance system to integrate data from laboratory reports, morbidity reports, medical records, and patient interviews. Additionally, an automated text messaging system was created by a multidisciplinary SNHD team that enabled public health authorities to communicate with new cases and contacts rapidly and reliably. This automated system also allowed SNHD disease investigators to submit case surveys and to share isolation and quarantine recommendations with individuals who tested positive for COVID-19.

Surging the workforce in Las Vegas

Authorities from the SNHD, state, and local governments worked to bolster the availability of the public health and health care workforces to be able to respond to the dramatic increase in COVID-19 cases and deliver the volume of health care services needed to respond to the COVID-19 pandemic. Primary support was provided by both state and local agencies assigned to emergency support functions, private sector EMS providers, other licensed medical providers, and agencies from partnering federal, state, tribal, and local governments. Additional personnel support was provided by the Las Vegas Visitors and Convention Authority, Clark County School District, local universities, hospitals, medical sector partners, and nongovernmental organizations.

The SNHD leveraged a combination of repurposing staff and contracting services through private providers and nonpaid volunteer organizations to surge the public health workforce to meet demand. Multiple SNHD programs redirected staff to support contact tracing and other social interventions. As a result, dozens of employees from other public health programs such as TB control, sexually transmitted disease, HIV/AIDS, and environmental health were used to support the pandemic response. The SNHD also relied on contracted agencies to provide hundreds of additional interviewers who delivered contract tracing services. The pandemic response also benefited from the contributions of community partners—such as the Nevada Medical Reserve Corps (MRC), Nevada National Guard, City of North Las Vegas Library, UNLV, and local businesses—who actively participated in the response and helped to expand the public health workforce.

The MRC was also used to surge the workforce. The MRC is a national program created after the 9/11 terrorist attacks to provide a mechanism to pre-register, train, and credential medical and other workers to deploy in the event of a disaster [19]. The entire MRC maintains a public health focus, and as a result, MRC units are generally housed in public health agencies, including the MRC of Southern Nevada. Through this program, MRC volunteers are periodically asked to participate in preparedness exercises and drills and are offered training. During the response to a disaster or infectious disease outbreak, such as the COVID-19 pandemic, MRC volunteers help the SNHD limit the spread or impact of the disease. MRCs who are medical personnel may volunteer for first aid and health roles, while nonmedical personnel may volunteer for other roles such as clerical, customer service, and distributing preparedness and health information. Some of the activities performed by volunteers throughout the pandemic have included augmenting medical and nonmedical staffing at the SNHD or medical facilities, assisting with the dispensing of medications, assisting with disease surveillance or notifications, assisting at staging areas to organize other volunteers, and serving in the community at first aid stations and health fairs. Throughout 2021, MRC volunteers in Clark County logged a total of 9197 total hours supporting the local COVID-19 response, translating to over $355,000 in savings that could be redirected to support other pandemic response services.

The MRC, however, is not the only opportunity for volunteers to bolster the public health workforce and those interested in helping in the pandemic response had many opportunities and organizations to choose from in Nevada. The State Emergency Registry of Volunteers-Nevada is a secure, web-based system used to register, qualify, and credential Nevada's health care professionals before a major public health or medical emergency. From this site, registrants may elect to join as a part of the aforementioned MRC, or a part of a statewide volunteer pool, or as a mental health crisis counselor. Additionally, the Governor of Nevada established the Battle Born Medical Corps as a part of his Emergency Disaster Declaration and Emergency Directives [20]. More specifically, in April 2020, the Governor activated the Nevada National Guard and signed the Battle Born Medical Corps emergency directive, which aimed at expanding Nevada's health care workforce to fight COVID-19 [21]. This directive waived certain licensing requirements to allow Nevada to quickly bring additional health care professionals (doctors, nurses, EMTs, and medical students) into hospitals and other health care facilities to care for COVID-19 patients.

Community partnerships were also key for planning and responding to the threats posed by COVID-19. The Clark County government was paramount in providing housing support, food, and transportation strategies. The SNHD partnered with the University of Las Vegas, Nevada to develop and maintain a contact tracing workforce. Private sector business partners—such as resort properties that maintained the space and logistical resources to service a high volume of people seeking testing and medical countermeasures—also aided in the pandemic response and the SNHD partnered closely with the resort and casino industry to find locations for clients to isolate and disclose negative test results with residents. Health authorities also worked closely with the Clark County School District, assisting them in the development of school mitigation guidelines, testing policies and protocols, including students and staff masking requirements. The SNHD also assigned a dedicated team of seasoned epidemiologists, disease investigators, and hygienists to offer technical assistance to the Clark County School District, including providing education to students, parents, teachers, and principals.

Ultimately, these efforts allowed for these critical workforces to expand at every level to successfully support community access to COVID-19 response services—whether it was testing and surveillance, contact tracing, establishing closed points of dispensing, or vaccinating the population once vaccines became available.

Surveillance and testing response: medical surge capacity and the need for increased surveillance and mass COVID-19 testing of the population in Clark County, Nevada

Throughout 2020 and 2021, health care and governmental agency responders provided significant community access to COVID-19 surveillance and testing services. On April 25, 2020, with support from North Las Vegas City Council members and Clark County School District nurses, the SNHD opened its first community-based collection site—a drive-through COVID-19 test collection site, located at the Martin Luther King Jr. Senior Community Center in the City of North Las Vegas. This site had 200 available tests per day, three drive-through testing lanes, and a tent for walk-ups. This event provided an important service that was in high demand, as all 200 tests were used within four hours of opening. In response to the increasing demand for COVID-19 diagnostic testing, additional mass testing sites and large venue drive-through test collection services were established at local high schools and at parking garages located at large event venues, such as casino properties, the Las Vegas Convention Center, the Las Vegas Cashman Center Complex, and others (Fig. 7.1). The experiences from standing up at these community testing sites also allowed responders to prepare for the future provision of mass vaccination operations.

Concurrently, the SNHD increased testing capacity at local hospital laboratories and contracted private labs, public health laboratories, and other organizations eligible to perform COVID-19 testing in Nevada. The increased demand for case investigation and contact tracing was addressed primarily through this expanded workforce and the use of contractors and staffing services.

FIGURE 7.1 A mass-testing drive-through site located in Las Vegas, Nevada.

Surging capacities to provide medical care and isolation services to those who did not meet hospital admission criteria or have a place to isolate

During the pandemic response, community partners and leaders recognized an urgent need to identify additional COVID-19 isolation locations and beds—beyond those provided by local hospitals and intensive care units—for patients who did not meet hospital admission criteria. Using local-level data, CDC modeling, and demand projections, community planners recommended that the SNHD identify an additional 1,000 beds capable of maintaining the infection control practices needed to safely isolate COVID-19-positive cases as a means of alleviating the demand for local hospital beds.

The convention center model that was used in several other states during the COVID-19 pandemic was not possible in Las Vegas, as shortages in the health care workforce meant that the necessary manpower was not available at the time. As a result, COVID-19 patients at these facilities would have been left without supportive care. Community leaders and planners used a variety of approaches to respond to these challenges. To begin, they looked elsewhere and decided to secure beds through a combination of facility types to achieve the goal of an additional 1,000 staffed beds. This included the use of rehabilitation hospitals that had available beds and staff.

They also sought to expand the workforce by leveraging contract staff, city and county fire EMS, and MRC volunteers. These individuals were used to staff venues such as the joint Clark County and City of Las Vegas Cashman Center Complex, which was used as a temporary isolation and quarantine facility that maintained the capacity to house up to 500 patients [22]. They were also deployed to contracted motels. Local governmental agencies worked to contract with motels that had been previously closed during the community-wide COVID-19-related shutdown, and secure vacant rooms to provide a location where care and support services could be delivered to confirmed COVID-19 cases that did not meet hospital admission criteria, had no place to stay in the county, and could care for themselves.

People experiencing homelessness also warranted special consideration. Homeless shelters isolated COVID-19-positive patients from their general population and worked with social service providers and hospital transfer centers to locate alternative isolation locations, such as rehabilitation and treatment facilities, for patients to recover. The SNHD also purchased a modular building to serve as an isolation and recovery site location for up to 30 COVID-19 patients. These efforts were supported through the CARES Act and FEMA Public Assistance funding, which allowed Clark County and other local governments to implement interventions focused on addressing the challenges posed by the unique needs of non-congregate shelters. At the time of writing, some of these services are still available in the county and have been very effective in addressing the needs of visitors and residents who need to comply with COVID-19 quarantine and isolation recommendations but lack housing options.

COVID-19 vaccination campaign operations

As the local health authority for Clark County and all the cities located within its boundaries, the SNHD had established partnerships well before the COVID-19 pandemic was declared a health emergency. These relationships were reinforced over time through the ICS's unified command during planned multijurisdictional exercises and through the response to other real-world events [13]. Still, the response to the COVID-19 pandemic highlighted the compelling need to expand partnerships to include additional organizations necessary for protecting the public's health and leading an effective community response. This need was most evident during mass vaccination campaigns for COVID-19.

With the arrival of COVID-19 vaccines in late 2020, the SNHD requested the support of the Clark County Type-3 Incident Management Team to aid in vaccination operations. These operations were informed by guidelines developed by the State of Nevada, Department of Behavioral and Public Health and included various segments of the population, as defined by priority and eligibility (Table 7.1). While the Clark County Department of Emergency Management worked very closely with the SNHD through the ICS unified command during the year 2020—supporting community-based, mass COVID-19 testing services and specimen collection operations—it soon became clear that the rollout of COVID-19 vaccines would require a separate ICS structure dedicated exclusively to vaccination efforts, as well

TABLE 7.1 Vaccination campaign in Clark County—timeline and target populations eligible to receive COVID-19 vaccinations.

Date implemented	Priority group and eligibility expansion
Dec 14, 2020	Tier group 1: health care workers
Jan 21, 2021	Cashman Convention Center Tier Group 1 (1st dose): health care workers, first responders, those 70 years of age or older
Feb 2, 2021	Cashman Convention Convention Center Tier Group 1 (2nd dose): health care workers, first responders, those 70 years of age or older
Feb 18, 2021	Vaccine eligibility expanded to 65 years of age or older
Mar 1, 2021	Eligibility expanded to all groups within frontline community support and frontline supply chain and logistics
Mar 10, 2021	Eligibility expanded to include transportation workers, air transportation staff, bus and public transit staff, and grocery store workers
Apr 5, 2021	Eligibility in Nevada opens to people 16 years of age and older
May 12, 2021	Pfizer vaccine offered to children 12–15 years of age
Nov 10, 2021	Pfizer vaccine offered to children 5–11 years of age
May 17, 2022	Pfizer booster doses available 5–11 years of age at all vaccine locations
Jun 18, 2022	CDC approves both Pfizer and Moderna for children 6 months–5 years of age
Jun 19, 2022	Eligibility expanded to include children ages 6 months–5 years of age
Jun 24, 2022	Moderna approved for kids 6 years–17 years of age

Source: Southern Nevada Health District, Office of Public Health Preparedness [23].

as additional personnel and resources to be able to run these concurrent operations. As a result, an additional ICS command team that supported vaccination operations was established under the same area command organizational structure.

On May 20, 2022, with the end of local and state disaster declarations in Nevada, and Clark County, the Type-3 Incident Management Team demobilized, and vaccination, surveillance, and testing operations transitioned back to a single ICS-managed structure. At the time of writing, the SNHD continues to provide these services to the public thanks to the federal disaster declaration and federal COVID-19 funding that supports ongoing COVID-19 response and mitigation activities.

Still, these actions were especially important in a context defined by limited resources. For instance, it was important to consider how to avoid reducing logistical support for COVID-19 testing while also establishing the new capacities required for vaccine distribution. To address this specific challenge, the SNHD testing efforts transitioned to rely more on contractors and resource staffing agencies to backfill personnel, while Nevada National Guard resources and EMS providers directly supported both testing operations, vaccine operations, and the distribution of pharmaceuticals, such as monoclonal antibodies (Fig. 7.2).

FIGURE 7.2   A Nevada National Guard disaster medical assistance team staffing a clinic providing monoclonal antibodies to COVID-19 patients.

The COVID-19 vaccination campaign in Clark County was also deployed through mega vaccination sites at Allegiant Stadium, the Cashman Convention Center, Texas Station Casino, the Las Vegas Convention Center, and the University of Las Vegas, Nevada. These, in addition to smaller community-based vaccination sites located at the SNHD main facility, public schools, churches, vaccination strike teams for homebound residents, and other community locations. As a result of these efforts, as of October 10, 2022, a total of 3,776,865 vaccination doses have been administered in Clark County—amounting to 74% of the population receiving at least one vaccination dose of COVID-19 vaccine, and 60% received at least two doses [24] (Fig. 7.3).

FIGURE 7.3   Members of the COVID-19 vaccination team offer vaccination services at Allegiant Stadium in Las Vegas.

Conclusion

In summary, despite the extreme limitations faced, especially during the early phases of the pandemic, the SNHD has led a relatively successful response to the COVID-19 pandemic. As discussed extensively throughout this chapter, the hiring and retention of health care personnel have been an immense challenge, which was partially addressed through repurposing the existing workforce, using contractor agencies, or outsourcing certain services to volunteer groups. This expanded workforce, overseen by the strong and cohesive leadership provided by the SNHD and the Clark County Regional Policy Group, has supported critical elements of the pandemic response including supporting testing and surveillance activities, providing services to patients at nontraditional health facilities, and vaccination efforts that have vaccinated over 1.7 million southern Nevadans.

Perhaps the most important lesson learned from this experience has been the importance of nurturing trust and multidisciplinary collaboration within the community. The experiences in Southern Nevada suggest that this can most effectively be done by actively engaging a large and diverse coalition of agencies, community-based organizations, and faith and industry representatives in the local response. Doing so works to create positive, open communication channels that allow for all segments of the community to actively contribute to the emergency response. In Clark County, this was successfully achieved, as evidenced by the multiple interventions held at local churches, community-based organizations, public schools, facilities, and mobile interventions in areas with large minority populations.

Unfortunately, despite these successes, the pandemic response has also been hampered by other broader forces and trends. Several important pandemic mitigation measures—such as the use of facemasks by the general population and the demand for life-saving COVID-19 vaccines—have been politicized and, as a result, become contentious. Relatedly, another notable challenge acutely felt in the pandemic response has been the interference of politicians in COVID-19 response operations, despite their lack of proper training and comprehension of public health concepts. Contributing to both of these challenges are obstacles presented by misinformation campaigns that rely on multiple communication channels, but especially social media, to amplify dangerous messaging and impede the implementation of protective COVID-19 measures. Because of this, as of October 10, 2022—and despite an extensive, integrated, and multisectoral response supported by local, state, and federal agencies—the COVID-19 pandemic has had a devastating impact in Southern Nevada. This includes over 605,000 confirmed COVID-19 cases, 30,150 hospitalizations, and more than 9,300 deaths.

This underscores an important reality that there is much work to do. There is a need to learn from this experience and to continuously plan for workforce surge, while strengthening relationships with community partners, local, state, and federal agencies to enable an effective response; there is a need for sustainable public health funding, especially when considering the growing risk of biological threats in our globalized world and the state of public health capacities in Nevada—a State that ranks at the bottom of the country when compared with other states' public health funding allocation per population; and there is a need to address the health inequities that have been exposed and exacerbated by the pandemic, especially among minority populations. Addressing these needs will help the SNHD and other local health authorities in Nevada improve their response to the COVID-19 pandemic and prepare for other biological threats of tomorrow.

References

[1] United States Census Bureau. QuickFacts: Clark County, Nevada. <https://www.census.gov/quickfacts/fact/table/NV,clarkcountynevada/PST045221>; 2022 [accessed 07.07.22].

[2] State of Nevada Department of Taxation. Population of Nevada's Counties and Incorp Cities <http://tax.nv.gov/Publications/Population_Statistics_and_Reports/>; 2022 [accessed 07.07.22].

[3] Applied Analysis. appliedanalysis.com. <http://www.appliedanalysis.com/work/data.php>; 2022 [accessed 07.07.22].

[4] United States Census Bureau. Census Business Builder Tool. <https://cbb.census.gov/rae/#view = report&clusterName = All + Sectors&industries = 00&geoId = 32003&geoType- = county&sectionsOff = undefined&reportType = detailed>; 2022 [accessed 07.07.22].

[5] Eadington WR. After the great recession: the future of casino gaming in America and Europe. Economic Affairs 2011;31:27–33. Available from: https://doi.org/10.1111/j.1468-0270.2010.02044.x.

[6] Las Vegas Convention and Visitors Authority. Visitor statistics. <https://www.lvcva.com/research/visitor-statistics/>; 2022 [accessed 10.10.22].

[7] Clark County Nevada, Southern Nevada Health District. Clark County recovery framework: multi jurisdictional COVID-19 return to hotel/resort operations, Operational Annex A. Las Vegas: Clark County Government; 2020.

[8] Southern Nevada Health District. Board of Health. <http://www.southernnevadahealthdistrict.org/about-us/board-of-health/>; 2022 [accessed 07.07.22].

[9] Department of Health and Human Services Nevada Division of Public and Behavioral Health (DPBH). Health Facilities. <https://dpbh.nv.gov/Reg/HealthFacilities/HealthFacilities_-_Home/>; 2022 [accessed 07.07.22].

[10] Packman J., Griswold T., Terpstra J., Warner J. Physician workforce in Nevada: a chartbook. Reno: University of Nevada, Reno School of Medicine; 2022.

[11] Nevada Legislature. Nevada Revised Statutes; Title 40 – Public health and safety; Chapter 439 – Administration of Public Health. Carson City: Nevada Legislature; 2022.

[12] Federal Emergency Management Agency. National response framework. Washington: United States Department of Homeland Security; 2019.

[13] United States Department of Homeland Security. Threat and Hazard Identification and Risk Assessment (THIRA) and Stakeholder Preparedness Review (SPR) Guide: Comprehensive Preparedness Guide (CPG) 201, 3rd edition. Washington: United States Department of Homeland Security; 2018.

[14] Cieslak PR, Noble SJ, Maxson DJ, Empey LC, Ravenholt O, Legarza G, et al. Hamburger-associated Escherichia coli O157:H7 infection in Las Vegas: a hidden epidemic. American Journal of Public Health. 1997 1997;87(2):176−80. Available from: https://doi.org/10.2105/ajph.87.2.176.

[15] Nguyen L. Salmonella I 4,5,12:i:- Gastroenteritis outbreak among patrons of firefly on paradise restaurant − Las Vegas, Nevada, 2013. Las Vegas: Southern Nevada Health District; 2014.

[16] Federal Emergency Management Agency. Developing and Maintaining Emergency Operations Plans Comprehensive Preparedness Guide (CPG) 101, Version 3.0. Washington: United States Department of Homeland Security; 2021.

[17] Federal Emergency Management Agency. Homeland Security Exercise and Evaluation Program (HSEEP). Washington: United States Department of Homeland Security; 2020.

[18] United States Centers for Disease Control and Prevention. Isolation and precautions for people with COVID-19. <https://www.cdc.gov/coronavirus/2019-ncov/your-health/isolation.html>; 2022 [accessed 08.19.22].

[19] U.S. Department of Health and Human Services, Administration for strategic preparedness & response. The Medical Reserve Corps. <https://aspr.hhs.gov/MRC/Pages/index.aspx>; 2022 [accessed 10.11.22].

[20] Government of the State of Nevada. Emergency Directive 011. Carson City: Government of the State of Nevada; 2020.

[21] Government of the State of Nevada. April 1 Press Release. <https://nvhealthresponse.nv.gov/wp-content/uploads/2020/04/4.1-Press-Release-Governor-Announcements_FINAL.pdf>; 2020 [accessed 10.11.22].

[22] City of Las Vegas. ISO-Q complex for homeless patients opens. <https://www.lasvegasnevada.gov/News/Blog/Detail/iso-q-complex-for-homeless-patients-opens>; 2020 [accessed 10.11.22].

[23] Southern Nevada Health District. Vaccination campaign in Clark County. <https://covid.southernnevadahealthdistrict.org/data/vaccination-campaign/>; 2022 [accessed 11.09.22].

[24] Southern Nevada Health District. COVID-19 case and vaccine data. <http://covid.southernnevadahealthdistrict.org/data/>; 2022 [accessed 08.19.22].

[25] United States Centers for Disease Control and Prevention. CDC updates and shortens recommended isolation and quarantine period for general population; 27 December 2021. Atlanta: US CDC.

Chapter 8

Using mobile financial services to improve community health workers' efficiency during the COVID-19 pandemic in Dhaka, Bangladesh

Farzana Misha[1], Syed Hassan Imtiaz[1], Margaret McConnell[2], Richard Cash[2] and Sabina Faiz Rashid[1]

[1]*BRAC James P Grant School of Public Health, BRAC University, Dhaka, Bangladesh,*
[2]*Department of Global Health and Population, Harvard T.H. Chan School of Public Health, Harvard University, Cambridge, MA, United States*

> *"What is a city if not the people."*
>
> — William Shakespeare (Coriolanus. 3.1.245)

Background

At the advent of the COVID-19 pandemic, urban areas were hit hard. According to a report published by the United Nations in July 2020, around 90% of all reported COVID-19 cases were from urban areas [1]. Inhabited by a population of 10.2 million individuals, Dhaka is one of the world's biggest megacities and also one of the most densely populated cities in the world [2].[1] The city is not only the capital of Bangladesh but also the country's most important political, economic, business, education, and health hub. Given this, over the past several decades, Dhaka has experienced an influx of low-skill economic migration from other parts of the country, resulting in a rapid rise in the urban poor population. This has subsequently resulted in increasing inequality in the city, overcrowding of transportation systems, and deteriorating living conditions, especially for the urban poor.

---

1. Megacities are defined as cities of 10 million inhabitants or greater.

Inoculating Cities, Volume II: Case Studies of the Urban Response to the COVID-19 Pandemic.
DOI: https://doi.org/10.1016/B978-0-443-18701-8.00009-6

The COVID-19 pandemic raised concerns and exposed the weaknesses of the existing health system and implementation strategies in Dhaka. It was also challenging for authorities to implement infection containment measures like social distancing and stay-at-home directives. However, compared to many developed countries, Bangladesh performed relatively well in terms of the rate of infection and casualties, which was primarily attributable to the rate of vaccination [3]. However, the journey was not smooth. With the help of the local government and city corporation, authorities tried implementing different interventions simultaneously in the city, making it difficult to identify which one of them was the most successful measure. Further, measures that were known to be effective faced significant implementation challenges. For instance, mass vaccination efforts are, without a doubt, one of the most effective response measures. However, raising awareness regarding the vaccines was a challenge for authorities amidst the mis- and disinformation widely spread over social media platforms.

In Bangladesh, these issues were further compounded by a fragmented urban health care system. While the Ministry of Health and Family Welfare (MoHFW) is responsible for health care in the country, the task of delivering these services in urban environments largely falls to the Ministry of Local Government, Rural Development and Co-operatives (MoLGRDC); and, as the MoLGRDC lacks the capacity necessary to provide such services, they frequently outsource tasks to nongovernmental organizations (NGOs) and private sector actors. This ultimately leads to a lack of coordination between service providers and the ineffective implementation of interventions. This *modus operandi* in the urban space leads to private health care services and NGOs playing a major role in the largely unregulated urban health care delivery system. There is also a stark dichotomy and inequity in the quality of care provided—with the rich accessing high-quality private health care facilities, and the poor relegated to receiving sub-par services from public health care providers. While the NGOs provide health care services geared toward the poor, they are mostly location and service-specific (i.e., antenatal care, vaccination, adolescent care, etc.). And, as a result, these services vary across locations, making it difficult to provide a blanket service to all urban poor.

This chapter focuses on the community health worker (CHW) model implemented by BRAC's Health, Nutrition and Population Programme (HNPP). Frontline health care workers such as CHWs in low- and middle-income countries are mostly involved in delivering essential health care services, communicating programmatic information, and communicating key health messages to the communities in a culturally sensitive manner [4]. The CHWs in the HNPP have provided door-to-door basic health care services to the urban poor since 2006. During the COVID-19 pandemic, these CHWs played a vital role in raising awareness, motivating individuals to get vaccinated, helping them with registration, accessing vaccination cards, and identifying nearby vaccination kiosks or centers. In the meantime, these CHWs continued to provide their usual health care services, including antenatal care, neonatal care, and other routine services.

Globally, most CHWs are women and are predominantly paid in cash. However, given the temporary and voluntary nature of their employment, ensuring a well-structured payment system is vital, as failure to do so may result in high rates of dropout and ineffective service delivery. To address these concerns, BRAC CHWs were introduced to mobile financial services (MFSs) in 2018. This was done primarily for administrative purposes, but evidence has demonstrated that access to MFSs also promoted financial inclusion (especially for women); improved CHWs' economic independence, savings, and access to remittances; acted as a mitigation strategy during emergencies; and even promoted empowerment [5,6].

In this chapter, we discuss the most effective strategies and interventions implemented by CHWs to contain the spread of the SARS-CoV-2 virus while also providing essential health care services to urban poor and slum populations. This chapter examines the strategies pursued in Dhaka to cope with the COVID-19 pandemic since its inception and attempts to explore the intricacies of how they were successful and can serve as models for responding to similar health emergencies in urban environments.

Background on Dhaka, Bangladesh

Dhaka is one of the oldest cities of Bangladesh situated next to the Buriganga River. Urbanized settlement in this city has been recorded as early as the 12th century [7]. In his book *A Geographical Account of Countries Round the Bay of Bengal*, East Indies merchant Thomas Bowrey describes Dhaka as a "large and spacious city" and admires it for its greatness, its magnificent buildings, and its multitude of inhabitants [8]. By Bowrey's account, Dhaka was "the largest city in Bengal and was famous for manufacturing cotton and silk." Indeed, some four centuries later, Dhaka is still admirable and has continued to function as a center of trade, commerce, and politics—though the city has undergone rapid development, the likes of which Bowrey could only imagine. Following the Partition of Bengal in 1947, Dhaka became the capital of East Pakistan and in 1971, the capital of independent Bangladesh.

Currently, Dhaka is one of the biggest megacities in the world [9]. The city is divided into two city corporations—the Dhaka North City Corporation (DNCC) and the Dhaka South City Corporation (DSCC), which bear populations of 5.9 million and 4.3 million, respectively—resulting in a total population of roughly 10.2 million. This represents an astounding 47% increase from the last census, which was conducted in 2011. The population density of Dhaka city is also the highest in Bangladesh, with 30,474 and 39,353 people per square kilometer in DNCC and DSCC, respectively; these estimates, compared to the national average population density, which is 2156 people per square kilometer [2].

The city, being the capital, is naturally the political center of the country—housing the Parliament, ministries, and numerous diplomatic institutions. It is also the financial and commercial capital of the country, accounting for 35% of the country's economy [10], as well as the digital hub of Bangladesh, as it has the highest percentage of mobile and internet users, across all genders. Overall, 79% of the population operates at least one mobile phone, and 48% of the Dhaka population uses the internet [2].[2]

Being the political, economic, and cultural hub of the country, the city has witnessed a 3.3% annual growth in population over the last few decades. Moreover, because Bangladesh is one of the most vulnerable countries with respect to climate change, the city is home to a substantial and growing number of climate migrants who move to the city, mostly in search of employment. This rapid urbanization and population growth are resulting in a dramatic increase in the urban slum population. Recent data estimate that the urban population comprises approximately 30% of the total population of the city, but this figure is expected to rise to 50% by the year 2050 [11,12]. According to the slum census of 2014, Dhaka has 3394 slums: 1639 slums in the DNCC that include 135,340 households, and 1755 slums in the DSCC with 40,591 households [13].

Dhaka is currently home to a total population of around 900,000 living in slums and floating settlements [2]. The city has struggled to provide basic services like housing, water supply, and education to this population. And, while urban environments can be more resilient in terms of health services and health infrastructure, when compared to rural environments, the rapidly increasing number of urban poor can strain these capacities and make it difficult to ensure equal and quality care, especially during a health emergency.

The health system in Dhaka, Bangladesh

Under Bangladesh's National Sustainable Development Strategy 2010−2021, urban development was identified as one of five strategic priority areas for sustainable development [14]; while the more recent 8th Five-Year Plan (FYP) 2021−2025 includes several strategies to be adopted as a means of improving urban conditions and reducing urban inequality [15]. Still, the extent to which these strategies and plans apply to the health system is unclear, as the health system in Bangladesh is one that is deeply fragmented, and oversight is interspersed between ministries.

More specifically, while the health system is governed by the MoHFW, the MoLGRDC plays a significant role in the oversight and provision of

---

2. Though gender imparities exist, with 58% of males reporting internet use, but only 39% of females reporting Internet use.

health care, especially as it relates to urban health. Furthermore, while urban public health centers are managed by local governments (i.e., city corporations and the local municipal governments), decision-making authority typically remains with the MoLGRDC and is rarely transferred to the local level. Thus, in Dhaka, acting under the MoLGRDC, the two city corporations supervise the health facilities (e.g., hospitals, clinics, and dispensaries), and also contract with NGOs and the private sector to provide health services. This includes the NGO Health Service Delivery Project and the Urban Primary Health Care Services Delivery Project. Other notable initiatives focused on providing health care services specifically to the urban poor include Marie Stopes Clinics and the BRAC Manoshi project.[3]

In partnership with the DNCC and DSCC, the MoLGRDC currently has seven and eight urban primary health care centers, respectively [16]. Though formal estimates made by the DNCC and DSCC state that the city has some 193 health facilities [17], these are underestimates and, in reality, more health care facilities exist in the city. The 8th FYP, however, highlights the previously referenced fragmentation issues, and the need to establish a permanent coordination structure between the two corporations to streamline the delivery of urban health and nutrition services in these facilities across Dhaka [15].

Despite these numerous facilities, there are still gaps and challenges in terms of public health coverage—especially for the urban poor and those residing in urban slums. The urban health system in Dhaka is dominated by private hospitals and clinics, where out-of-pocket expenses are high (72.68% in 2019) and often prohibitive for the urban poor [18]. Through partnerships among the government, development partners, and NGOs, the past two decades have witnessed efforts to improve the provision of primary health care in urban areas [19]. However, it is not adequate in its current form, and evidence shows that the urban poor primarily rely on local pharmacies, homeopaths, and traditional healers due to a lack of equitable access to quality health care [20].

One strategy is to improve service delivery at the community level through CHW models and programs. Evidence has demonstrated that CHW programs have the potential to reach marginalized populations, and a meta-analysis investigating the impacts of task shifting to CHWs found that CHWs could safely and effectively deploy interventions for various communicable and noncommunicable diseases and that such programs could result in substantial cost savings [21,22]. Accordingly, many NGOs have integrated CHWs into their health interventions. For example, in 2007, BRAC initiated

---

3. The NGO Health Service Delivery Project has supported the delivery of primary health care through a nationwide network called *Surjer Hashi*, or "Smiling Sun"; network consists of more than 25 local NGOs, 350 + static clinics, and 10,000 + satellite clinics. Marie Stopes, on the other hand, is an NGO that has provided SRHR services in Bangladesh since 1988.

its Manoshi community health care program to provide essential pre- and post-natal health services in the urban slums of 11 city corporations, including BNCC and BSCC [23].

Other efforts for improving the health system in Bangladesh include digital health initiatives. Despite the hurdles existing in the country's health system, one major area of growth has been its immense progress in the digital health sector. These efforts began in 2009 when the current government started promoting health digitization as one of its key objectives, with the slogan, "Digital Bangladesh." This objective has included the deployment of the open-source District Health Information Software 2, which has enabled the health system to maintain a health data warehouse that enables data-driven decision-making [24]. Other efforts have included using bulk SMS messaging, which has been used since 2009 to provide health information on immunization, noncommunicable diseases, infectious diseases, antibiotic resistance, maternal health, and other health priorities. Additionally, the Directorate General of Health Services has started to use social media platforms, such as Facebook, to broadly disseminate important health information [25].

Nevertheless, in terms of policies, notable challenges still exist. The Medical Practice and Private Clinics and Laboratories (Regulation) Ordinance of 1982, the foundational legal text for health data in Bangladesh, does not contain a provision for sharing health data with the government [26]. Thus, while most private hospitals and health facilities have well-developed data storage systems in place, they are not obligated to share health information with the government. There is some precedent for the sharing of this data—as, during a dengue outbreak in the 2000s, private hospitals were asked to share information on dengue and other infectious diseases with the MoHFW—but also a clear need for this ordinance to be updated with regard to digital health infrastructure and data governance.

COVID-19 in Dhaka, Bangladesh

The first case of COVID-19 was reported in Bangladesh on March 8, 2020, as confirmed by the Institute of Epidemiology, Disease Control and Research, and the country observed its first death due to COVID-19 on March 18, 2020 [27]. Due to the rising case counts in the following weeks, the country reduced international flights, imposed thermal scanner checks at airports, and shut down all educational institutions by mid-March. It also introduced a mandatory 14-day quarantine for all foreign travelers entering the country [28].

Despite the numerous initiatives to respond to the COVID-19 pandemic, discussed in greater detail later in this chapter, Dhaka experienced the highest number of cases and deaths in Bangladesh [29]. While Dhaka city maintains the highest literacy rates among all districts in the country, it is also the most inequitable in terms of socioeconomic status, where disparities prevent the poor from accessing basic services, such as health care and education. Because of this, the advent of the COVID-19 pandemic was a notable concern, especially for the urban poor in Dhaka. Previous work has discussed the impacts of disasters in urban areas, suggesting a variety of necessary measures and potential recovery plans [30]. However, because pandemics are relatively rare, a more limited literature exists on the impacts of pandemics on cities [31]. The literature that does exist has emphasized the inequalities faced by marginalized and vulnerable groups, with noted examples found during the 14th-century plague pandemic (i.e., the Black Death), and during a smallpox epidemic in the 1850s [32]. In a megacity like Dhaka, the dense population and the high level of interconnectivity at both global and local scales make this population even more vulnerable to infectious disease outbreaks.

This is especially true as the COVID-19 pandemic exacerbated poor living conditions and health risks that existed before the pandemic. In addition, the urban poor also had to grapple with added uncertainties regarding job security, as well as income loss resulting from restricted mobility interventions that were applied during the pandemic [33]. In such instances, it was necessary to have a well-established urban health system that could easily provide all types of health care delivery services to city dwellers.

Response measures adopted in Dhaka

Bangladesh, like most countries, adopted nonpharmaceutical interventions as a first tool to control the spread of the COVID-19 virus, before supplementing these with clinical management. As COVID-19 case counts rose, Bangladesh adopted measures like social distancing and partial or complete lockdown. However, these interventions lacked adequate planning. On March 25, 2022, the country announced its first lockdown, which was scheduled to last for 10 days. In response to this announcement, millions of people sought to leave Dhaka, using any means possible to travel to their hometowns, often violating the social distancing instructions. And, before the 10-day period was over, the city experienced an influx of thousands of people coming back to the city for work, primarily due to a lack of coordination between industry (i.e., the ready-made garment sector) and the government. Even though the lockdown measures were further extended, the effectiveness of these initial measures has been subject to scrutiny, and it seems likely that these actions backfired in their efforts to reduce the infection rate due to inadequate planning.

The government also initiated awareness-raising campaigns that primarily utilized traditional media sources, such as newspapers, television, and radio. However, this approach mostly precluded the poor, who did not necessarily have the means to access this information. In parallel, while the urban poor are generally viewed as being well-connected through social media, these platforms were rife with mis- and disinformation throughout the pandemic response, which complicated these efforts. Thus, the narratives surrounding the pandemic have been shaped and strongly influenced by information that is not always accurate—contributing to misconceptions that have resulted in fear, shame, and stigma [34]. The government used a variety of digital initiatives to attempt to address these challenges. These included partnering with mobile network providers to use awareness-raising phone recordings that were played before calls, sending informational text messages, and posting information on the social media profiles of various government ministries. Other community-oriented initiatives included partnering with local leaders, such as religious leaders, to disseminate information in speeches at mosques and other places of worship.

Many people in Dhaka use public transportation systems, including buses and trains, to commute, and these systems are often overcrowded. Because of this, in efforts to ensure the proper implementation of social distancing measures, the government mandated passenger reductions on public transportation systems and advised that systems should reduce the number of passengers by half. However, these mandates were widely criticized by both civil society and transportation companies, most directly impacting wage workers, who constitute a majority of the urban poor, which made their enforcement challenging [35]. As a result, it was eventually concluded that strictly enforcing protocols like social distancing and lockdown measures was impractical in a densely populated megacity.

Clinical measures for treating confirmed COVID-19 cases were developed by the Directorate General of Health Services (DGHS) and published as the National Guidelines on Clinical Management of COVID-19 [34]. These guidelines included a range of considerations, such as treatment protocols, management strategies, post-covid care, and vaccinations, and were updated several times as more information was learned about COVID-19 and the World Health Organization updated its guidelines and recommendations.

In February 2020, the Kuwait Maitree Hospital in Dhaka was designated by the government as a dedicated COVID-19 hospital. Later, in May 2020, all private and public hospitals in the city were instructed to treat COVID-19 patients. Capitalizing on the existing digital ecosystem, information on the availability of hospitable beds, intensive care unit beds, and ventilators was made available online. This allowed individuals to monitor the situation and for patients to target hospitals and plan accordingly in case of emergencies.

The DGHS also worked with the private sector to support the response. Several private entities were assigned responsibility for conducting COVID-19 diagnostic tests and delivering results. Private entities were also assigned to supply medical protection equipment and products (i.e., personal protective equipment, masks, etc.). However, there were several scams and allegations of corruption regarding these assignments [36]. Following these allegations, the government acted quickly, as the Anti-Corruption Commission submitted a charge sheet against the DGHS, which contributed to several resignations amidst heavy criticism.

Once COVID-19 vaccines had been developed, the government also developed and implemented a successful strategy to work toward ensuring vaccination. Since early 2021, Bangladesh began to procure vaccines from various sources. Between January and April 2021, only the Oxford-AstraZeneca vaccine was authorized for use in the country. Doses of CoviShield from India were ordered for use during this time, but the supply was halted as a result of India's surge in cases. The country suspended its vaccination program in April but later renewed these efforts in earnest after approving several additional vaccines (i.e., Russia's Sputnik V, China's BBIBP-CorV, Pfizer-BioNTech, and later Moderna). Bangladesh received supplies of the Pfizer, AstraZeneca, and Moderna vaccines through COVAX, and other vaccine supplies were received through donations from China, the United States, Japan, and India.

These efforts were supported through the development and use of a digital app. On January 27, 2021, the government launched the Surokkha app—its first-ever digital app for registering for mass vaccination—and mass vaccination campaigns began the following week.[4] The app was designed to allow individuals to register for vaccination at designated health facilities and provide users with vaccination cards and certificates [37]. This system was used to help inoculate the population against COVID-19, and as of August 25, 2022, some 292,888,900 doses of COVID-19 vaccine have been administered in Bangladesh—equating to nearly 72.3% of the population completing a two-dose vaccination regimen [38,39]. This percentage is higher in the capital district, with some 84.2% of the population in Dhaka receiving their second dose of vaccine. Beginning in February 2021, the government prioritized vaccinating the elderly over the age of 77, with succeeding age groups; at the time of writing, the government is prioritizing vaccinating school-aged children, primarily through school-based vaccination campaigns.

The government also took steps to provide social safety nets to the poor. As outlined in the National Social Security Strategy, the government set up systems to identify the regions across the country (both urban and rural) most affected by the crisis so that payments to beneficiaries could

---

4. Initially, Bangladesh purchased the Oxford-Astra-Zeneca vaccines produced by the Serum Institute of India. Later, the government approved and provided the Russian Sputnik V, and the Chinese Sinopharm BIBP vaccines. In mid-2021, under the COVID-19 Vaccines Global Access (COVAX), Bangladesh received Pfizer's, Moderna, and Sinopharm BIBP vaccines. Moreover, in 2020, a Bangladeshi pharmaceutical company, Globe Biotech Limited, declared intentions to locally develop a COVID-19 vaccine named Bangavax and sought permission for conducting clinical trials. The government approved these efforts but after 1 year, the trials had not been conducted, nor authorized by Bangladesh Medical Research Council.

be increased, which were supported by action plans developed by the MoLGRDC, the Ministry of Disaster Management and Relief, and other pertinent ministries [40]. On May 14, 2020, the government announced plans to provide support to 5 million low-income families impacted by the pandemic in a one-time government-to-person cash transfer of 2500 BDT. The Ministry of Disaster Management and Relief contributed to this effort by preparing a list of the people based on information collected by the district commissioners, Upazila Nirbahi Officer offices, and city corporations across the country.[5] The list included rickshaw pullers, van drivers, day laborers, construction workers, agricultural laborers, transport workers, hawkers, restaurant employees, and domestic helpers, among others [41]. Cash transfers were then provided through major MFS providers. However, there were some concerns raised by several organizations and civil society that involved poor governance of this disbursement, technical complexity (i.e., planning, implementation, and insufficient coverage), and the procedures used to identify the individuals most in need of cash transfers.

From the onset of the pandemic, several NGOs and voluntary workers also vigilantly participated in raising awareness, providing COVID-19 safety gear, and providing health services to the most vulnerable populations in urban areas. Among these is BRAC, which continued to provide health care services through its health care centers established under the BRAC HNPP. Within the BRAC HNPP, the flagship CHW program also played a significant role in the pandemic response, as rapid community engagement and service delivery were only possible by using an established system—one that had proper reporting mechanisms and had been sufficiently digitized. In the next section, we will discuss how the BRAC CHW model was established, how it was scaled-up and partially digitized prior to the pandemic, and how the model was used during the COVID-19 pandemic to provide services to and support the urban poor.

Community health workers and mobile financial services

One of the recurring challenges of providing health care in developing countries is the disconnect between the poor and formal health care providers. The situation in Dhaka is no different, and the health care landscape for the urban poor population is bleak. Due to ambiguity regarding precise roles and insufficient planning and coordination in the existing health system, slum dwellers are left with inadequate access to basic health care services [42,43]. This situation is complicated by the reality that this population is disproportionately impacted by forces like social fragmentation, environmental

---

5. Upazila Nirbahi Officers are the chief executive officer of an upazila—an administrative region that may be viewed as analogous to a county or borough in other governance contexts.

hazards, and transient living conditions, which can be coupled with crime and violence, ultimately rendering them more susceptible to shocks (i.e., economic, health, political, etc.) [11,12]. Accordingly, establishing equitable and quality health services for this segment of Dhaka's population has become a pressing issue.

Evidence shows that, in low-resource settings, CHWs can build bridges between formal health systems and communities and improve the accessibility and acceptability of health services [44]. Therefore, CHW models are widely regarded as an effective health care delivery strategy, especially for underserved populations. Additional evidence has demonstrated that CHWs can provide basic health care services and ensure access to higher levels of care by referring those in poorer households [45,46]. Beyond these immediate benefits, CHW models can also provide positive byproducts, such as empowering the women who often work in these roles [47,48].

BRAC's journey with CHWs started in 1976 through the formation of a cadre of Lady Family Planning Organizers. Currently, the CHW model consists of two cadres—Shastho Shebika (SS), who function as voluntary health care workers, and Shastho Kormi (SK), who function as salaried health care workers. The program maintains a gendered perspective and focuses on using female health workers in Bangladesh to address sociocultural barriers to accessing health care and providing health services.

Currently, a total of 1716 CHWs (156 SK CHWs and 1560 SS CHWs) work in Dhaka under the 20 BRAC medical centers.[6] These CHWs receive three weeks of training and spend between two and four hours per day, six days a week, delivering health care services. This typically allows for CHWs to visit between 10 and 30 families per day. Their primary responsibility is to disseminate information on health, nutrition, and family planning as well as encourage families to adhere to safe water, sanitation, and hygiene practices. Another responsibility of CHWs is to identify pregnant women in their community, explain the importance of antenatal care, and refer them to government health centers or BRAC health clinics.

The introduction of mobile financial services to BRAC community health workers and health service delivery

Mobile financial services have been found to increase consumption, savings, labor outcomes, remittances, and migration [49–53], and studies have also shown that digitization of CHW payments can improve coverage and quality care [6]. The use of MFSs in health financing, however, remains relatively limited.

---

6. BRAC has a total of 40 medical centers across the country of Bangladesh, 20 of which are located in Dhaka.

In the late 2010s, one concern of the BRAC HNPP was notable delays in the submission of collected fees by the CHWs following service delivery. Although CHWs were advised to submit collected fees on a bi-weekly basis, they were more likely to submit fees at the end of the month, during their monthly visits to the branch offices. Upon investigation, HNPP management found two possible reasons for these actions. One, involved CHWs avoiding travel to the branches on a bi-weekly basis because it was inconvenient to do so solely for submitting fees; the other possible reason involved CHWs using the collected cash during the month for immediate consumption and then submitting fees at the end of the month from their own savings. Irrespective of the reason, as a result, it was difficult for BRAC staff to monitor service delivery throughout the month. It also made it challenging to make changes, should they be required (e.g., increasing certain types of services, etc.). Because of this, recognizing the potential offered by MFSs and in efforts to improve operations and collect the money immediately, the HNPP made the administrative decision to introduce MFSs in 2018.

More specifically, the introduction of MFSs was rationalized to enable the prompt collection of fees and to reduce the logistics, travel time, and costs associated with CHWs visiting branch offices to deposit collected fees. With these objectives, three programs of BRAC—the Social Innovation Lab, HNPP, and the Microfinance Program—collaborated to share knowledge and provide logistic support for a three-month pilot program where CHWs used MFSs. Following the success of this program, the initiative was scaled-up to all 4300 CHWs across the BRAC branches.

Currently, the BRAC CHWs use the MFS bKash to send the service charges received from the beneficiaries of health services to BRAC. Under this model, CHWs use specific reference codes to send money to designated merchant accounts for services delivered. When CHWs send money, the transaction history is updated on the MFS server, which can be accessed and reviewed by BRAC operational staff. When CHWs deliver a service package, they also provide beneficiaries with an official receipt for validation purposes. BRAC area offices also keep receipts of service packages. These records are digitally maintained in area offices using tablets, which replaced register books that were used for cross-checking before MFSs were rolled out.

While the benefits of MFSs were undeniable, the decision to scale-up the model brought new challenges with it. From a logistical perspective, the scale-up required significant server and software support from BRAC Information & Technology Services. BRAC University's Information and Technology Department also assisted in these processes by generating digital wallets that CHWs could use to deposit money. Beyond these, additional challenges are presented in field operations. These were mostly associated with the acceptability and use of MFSs by CHWs and BRAC area managers. When MFSs were initially introduced, concerns existed regarding whether

CHWs—who predominantly come from poorer backgrounds with low literacy rates—would be able to effectively adopt the digital technology. Other issues related to CHWs requiring assistance from their husbands or other male family members to use the digital services and access MFS accounts. To address these issues, BRAC provided CHWs and area managers with numerous training sessions to instruct them on how to use MFSs. Language barriers were also a concern, as the bKash app initially only offered English language options, but this was later addressed by the app when Bangla was introduced as an additional language option.

Despite training, some challenges persisted. Most notably, CHWs found it challenging to correctly enter the reference numbers, deposit the correct amount of money, register a valid phone number/bKash wallet number, and recognize fraudulent calls. To mitigate these challenges, regular refresher training sessions were held, and greater assistance was offered to CHWs by establishing a new team at the BRAC branch offices.

Community health workers and the use of mobile financial services during the COVID-19 pandemic

During the pandemic, when lockdown measures were applied throughout Bangladesh, the BRAC CHWs were still providing health services to the urban poor, and having access to MFSs helped them continue providing these services with fewer setbacks. When the pandemic began, BRAC HNPP's first response was to raise awareness about COVID-19 using CHWs. The SK and SS CHWs were provided with informational leaflets and posters to raise awareness in the communities about important response measures, like social distancing. In April 2020, in a massive risk communication campaign that sought to provide health information related to COVID-19, BRAC instructed 50,000 CHWs all over the country, to visit at least 10 homes for 10 days—providing some 5 million households with health and hygiene information, stickers, and other important updates.

In Dhaka, the BRAC CHWs operated in 68 wards to support the Community Support Team Project, in collaboration with the United Nations Population Fund and the Food and Agriculture Organization, and with financial support provided by the United Kingdom's Foreign, Commonwealth, and Development Office. As a part of this project, CHWs distributed free masks, ran hand washing booths, supported vaccine registrations, arranged telemedicine services, and conducted community health monitoring—all in addition to the more routine health services they provided to the urban poor. These efforts were notable as they filled a void in data. Initially, the Institute of Epidemiology, Disease Control and Research and the DGHS gathered data on the number of tests performed, the number of positive cases and fatalities. This information was published in an online DGHS dashboard that provided day-to-day updates regarding the situation. But these raw data were not publicly available, and they were updated less frequently once rapid antigen testing became widely available. This is where the CHWs' efforts played a vital and catalytic role in their respective communities, as their work helped to provide information on basic health services provided, symptoms, tests conducted (including rapid antigen testing), and other health priorities.

The CHWs were also heavily engaged in building awareness regarding COVID-19 infection. As mentioned earlier, misinformation and some media coverage created misconceptions and stigma that further complicated response efforts. The CHWs played a vital role in addressing these issues [54]. Despite the fear and anxiety surrounding the novel virus, CHWs were welcomed in their respective communities throughout the pandemic, and interviews with the CHWs suggested that the communities they served received their door-to-door visits positively. There were some instances in which CHWs were requested not to enter a household, often by individuals who sought to mitigate the risk of exposure to the virus (i.e., since CHWs were visiting every household and were exposed to many individuals). Still, their work in raising awareness around COVID-19 was appreciated by community members, and findings show that greater than 70% were praised by the beneficiaries for their services amidst the pandemic [55]. This was possible, largely because of their pre-established and trusted relationships with the communities they served and because they are a part of the community they provide services to.

Experiences and challenges in providing financial support to community health workers through mobile financial services

The imposed movement restrictions during the initial phases of the COVID-19 pandemic reduced the number of door-to-door visits, resulting in a reduction in the SK CHWs' performance bonuses and the SS CHWs' income from selling basic health and hygiene products to the households. To offset the financial burden and for talking up extra workload for raising awareness, the BRAC HNPP provided CHWs with a two-stage financial incentive through bKash. CHWs initially received a 1000 BDT cash incentive, which was later supplemented with an additional incentive of 1500 BDT as the work continued.[7]

At the beginning of these efforts, BRAC's Social Innovation Lab collaborated with bKash to confirm the feasibility of such a large amount of disbursement. Moreover, they had to identify the appropriate mobile wallets to disburse funding. Of the 50,000 health workers targeted for cash assistance by this program, nearly 22,000 had mobile wallets ready for transfers. Almost all of the SK CHWs had a personal mobile wallet registered in their

---

7. The additional incentive was provided by the United Nations Office for Project Services that helped to support various projects around the world.

own name, but most of the SS CHWs provided the mobile wallet account numbers of either their husbands or other family members, and some were reluctant to open new accounts. To address this, HNPP reached out to around 43,000 SS CHWs to help them complete their own registration. In the beginning, the HNPP planned to bring the SS CHWs to a center to complete the registration process, but the movement restrictions imposed throughout Bangladesh made this infeasible. After consulting with bKash, HNPP asked the area managers to collect relevant information from the CHWs—such as their national identity card number, mobile number, whether they had their own mobile wallet, and their bKash account number if they did have their own mobile wallet. The HNPP then asked the SS CHWs without an account to open their own personal account and share that account number with their area manager. With this information, and with the help of BRAC Information & Technology Services, the HNPP created a database of 40,000 SS CHWs and their account numbers, which was subsequently verified by bKash. The entire verification process was lengthy, as the HNPP revised the information constantly to detect and delete any incorrect information. But this represented a significant development, as it marked the first time that SS CHWs were integrated into the MFS system, and the HNPP was able to reach approximately 60% of the SS CHWs using bKash.[8]

Time cost efficiencies due to mobile financial services

As briefly discussed earlier, the use of MFSs can improve the health services provided by CHWs because they simplify logistics and reduce administrative travel costs, which allow CHWs to prioritize providing services to hard-to-reach communities and seek out patients who are the targets of stigma and discrimination. Internal data from a BRAC HNPP survey suggests that MFSs allowed the SK CHWs and SS CHWs to save 6.7 and 3.4 hours per month, respectively, where time was saved primarily on money transactions. It is estimated that this time saved allowed CHWs to spend 20 additional minutes with patients per day, or two minutes per household visited (i.e., assuming they visit 10 households per day). Moreover, apart from saving time, MFSs also resulted in a 15% reduction in SK CHWs errors associated with submitting health care service fees following the introduction of MFS. While the referenced data included CHWs in both urban and rural environments, the broader implications of these results are still attractive from a policy perspective and suggest that MFSs can save both time and money during the response to health emergencies.

The survey also asked CHWs how MFSs could play a role during a crisis—whether it be a pandemic, flood, or other natural disaster. The responses

---

8. The HNPP invited the remaining SS CHWs without mobile wallets and account numbers to visit their nearest bank branch to collect their cash incentives.

received were varied but contained several broad themes. Included were economic incentives, which referred to the stimulus package provided to them by BRAC, government benefits, and the social safety nets provided amidst lockdown measures; emergency health services, which referred to rapid financial transactions using MFSs in the event of emergencies and empowering CHWs to purchase necessary items, especially when traveling in person was not possible. CHWs also suggested that MFSs could provide benefits in times of crisis beyond the health sector. For instance, educational services referred to the ability of students to receive stipends instantly and of teachers to be paid using MFS for providing online classes or support to the students

Conclusion

Despite the numerous initiatives in Bangladesh to respond to the COVID-19 pandemic, the outbreak exposed weaknesses in Dhaka city's health infrastructure and, when compared to other districts in Bangladesh, the city experienced the highest case and death counts in the country. This was, in part, due to the notable inequalities that persist in the city, as the pandemic highlighted areas that need to be prioritized in the future to deliver more effective services during health emergencies.

The positioning of CHWs within the urban health system—where they have door-to-door access and are well-regarded in the communities they serve—helped to blunt the impacts of the pandemic on the urban poor. Because this model had been championed in Dhaka before the pandemic began, CHWs had an upper hand in raising awareness at the community level while also continuing their critically important work of providing more routine health services. The CHWs also helped to collect important data on the state of health in the city's most vulnerable populations throughout the public health emergency.

Importantly, the use of MFSs facilitated this data collection and CHW operations. Because of MFSs, it was possible for the BRAC HNPP to continue providing services, collect information on service delivery, collect service fees, and provide financial support to the CHWs throughout the pandemic.

Ultimately, the experiences from Dhaka suggest that CHWs can work as important conduits for governments to contact hard-to-reach populations in urban areas and provide them with health services during public health emergencies. These models can be augmented using basic digital services, like MFSs, which can connect CHWs to existing digital data collection systems and ensure digital financial inclusion and hassle-free financial transactions. Hence, future efforts that seek to build upon CHW models and improve pandemic preparedness should ensure that the CHW networks are established and operating efficiently before the response to an actual emergency, especially as a means of promoting equitable responses and reaching the urban poor.

References

[1]   United Nations. Policy Brief: COVID-19 in an Urban World. New York: United Nations;; 2020.

[2]   Bangladesh Bureau of Statistics. preliminary report on population and housing census 2022. Dhaka: Bangladesh Bureau of Statistics; 2022.

[3]   Our World in Data. Coronavirus (COVID-19) vaccinations, <https://ourworldindata.org/covid-vaccinations>; 2022 [accessed 27.06.22].

[4]   McConnell M, Mahajan M, Bauhoff S, Croke K, Verguet S, Castro M, et al. How are health workers paid and does it matter? Conceptualising the potential implications of digitising health. BMJ Global Health. 2022;7:e007344. Available from: https://doi.org/10.1136/bmjgh-2021-007344.

[5]   Kim K. Assessing the impact of mobile money on improving the financial inclusion of Nairobi women. Journal of Gender Studies 2022;31(3):306−22. Available from: https://doi.org/10.1080/09589236.2021.1884536.

[6]   Garz S, Heath R, Kipchumba E, Sulaiman M. White paper: evidence of digital financial services impacting women's economic empowerment. Dhaka: BRAC Institute of Governance and Development; 2020.

[7]   Ahmed SU. Dacca: a study in urban history and development. London: Routledge; 1986.

[8]   Bowrey T. A geographical account of countries round the Bay of Bengal. Cambridge: The Hakluyt Society; 1905.

[9]   Ahmed K., Montu R.I. The making of a megacity: how Dhaka transformed in 50 years of Bangladesh. *The Guardian*; 26 Mar 2021.

[10] Ahsan A. Dhaka centric-growth: at what cost? Policy Insights; 21 Nov 2019.

[11] Ahmed I. Factors in building resilience in urban slums of Dhaka, Bangladesh. Procedia Economics and Finance 2014;18:745−53. Available from: https://doi.org/10.1016/S2212-5671(14)00998-8.

[12] Ahmed I. building resilience of urban slums in Dhaka, Bangladesh. Procedia Social and Behavioral Sciences 2016;218:202−13. Available from: https://doi.org/10.1016/j.sbspro.2016.04.023.

[13] Bangladesh Bureau of Statistics. 2014 Census of slum areas and floating populations. Dhaka: Bangladesh Bureau of Statistics; 2015.

[14] Bangladesh Planning Commission. National sustainable development strategy (2010−2021). Dhaka: Government of the People's Republic of Bangladesh; 2013.

[15] General Economics Division. 8th Five Year Plan: July 2020−June 2025. Dhaka: Government of the People's Republic of Bangladesh; 2020.

[16] Ministry of Local Government Rural Development & Co-Operatives. Urban primary health care centers, <http://uphcsdp.gov.bd/city-corporations>; 2019 [accessed 27.06.22].

[17] Dhaka South City Corporation. Health facilities in Dhaka South City Corporation, <http://dscc.gov.bd/site/page/c6818c61-93b4-4b91-97c9-ab814113a88b/->; 2022 [accessed 27.06.22].

[18] World Bank. Out-of-pocket expenditure in Bangladesh, <https://data.worldbank.org/indicator/SH.XPD.OOPC.CH.ZS?locations = BD>; 2022 [accessed 27.06.22].

[19] Ahmad A. Provision of primary health care services in urban areas of Bangladesh: the case of urban primary health care project. Department of Economics Working Paper No. 9. Lund: Lund University; 2007.

[20] Adams AM, Islam R, Ahmed T. Who serves the urban poor? A geospatial and descriptive analysis of health services in slum settlements in Dhaka, Bangladesh. Health Policy and Planning 2015;30(1):i32−45. Available from: https://doi.org/10.1093/heapol/czu094.

[21] Ahmed S, Chase LE, Wagnild J, Akhter N, Sturridge S, Clarke A, et al. Community health workers and health equity in low- and middle-income countries: systematic review and recommendations for policy and practice. International Journal for Equity in Health 2022;21:49. Available from: https://doi.org/10.1186/s12939-021-01615-y.

[22] Seidman G, Atun R. Does task shifting yield cost savings and improve efficiency for health systems? A systematic review of evidence from low-income and middle-income countries. Human Resources for Health 2017;15(29). Available from: https://doi.org/10.1186/s12960-017-0200-9.

[23] BRAC. Manoshi Program of BRAC, <http://www.brac.net/program/health-nutrition-and-population/maternal-neonatal-and-child-health/manoshi/>; 2022 [accessed 27.06.22].

[24] Khan MA, Azad AK, Cruz VD. Bangladesh's digital health journey: reflections on a decade of quiet revolution. WHO South-East Asia Journal of Public Health 2019;8 (2):71−6. Available from: https://doi.org/10.4103/2224-3151.264849.

[25] Waldman L, Ahmed T, Scott N, Akter S, Standing H, Rasheed S. We have the internet in our hands': Bangladeshi college students' use of ICTs for health information. Globalization and Health 2018;14(31). Available from: https://doi.org/10.1186/s12992-018-0349-6.

[26] Ministry of Law, Justice and Parliamentary Affairs. Ordinance No. IV of 1982 − the medical practice and private clinics and laboratories (regulation) ordinance. Dhaka: Ministry of Law, Justice and Parliamentary Affairs; 1982.

[27] Reuters. Bangladesh reports first coronavirus death - officials. Reuters; 18 Mar 2020.

[28] Maswood M., Chowdhury S. Bangladesh Bans Travelers' Entry From Europe. New Age, 14 Mar 2020.

[29] Directorate General of Health Services. COVID-19 dynamic dashboard for Bangladesh, <http://dashboard.dghs.gov.bd/webportal/pages/covid19.php>; 2022 [accessed 27.06.22].

[30] Sharifi A, Khavarian-Garmsir AR. The COVID-19 pandemic: impacts on cities and major lessons for urban planning, design, and management. Science of the Total Environmnet 2020;749:142391. Available from: https://doi.org/10.1016/j.scitotenv.2020.142391.

[31] Matthew RA, McDonald B. Cities under siege: urban planning and the threat of infectious disease. Journal of the American Planning Association 2006;72(1):109−17. Available from: https://doi.org/10.1080/01944360608976728.

[32] Wade L. An unequal blow. Science (New York, N.Y.) 2020;368(6492):700−3. Available from: https://doi.org/10.1126/science.368.6492.700.

[33] Ahmed Raha S, Rana S, Al Mamun S, Hasan Anik M, Roy P, Alam F, et al. Listening to young people's voices under covid-19 − revisiting the impact of covid-19 on adolescents in urban slums in Dhaka, Bangladesh: Round 2. London: Gender and Adolescence: Global Evidence (GAGE) Programme; 2021.

[34] Siddiqui S, Nowshin N. "They Won't Even Touch the Money We Touched": Stigma, Shame and COVID-19. Dhaka: BRAC James P Grant School of Public Health, BRAC University; 2020.

[35] Khatun F. Surviving through the second lockdown. The Daily Star; 5 Apr 2021.

[36] New Age. Ex-health DG, 5 others to face Regent hospital scam charge. New Age; 21 Sep 2021.

[37] Ahamed S., Masum O. Limited vaccine registration only on Surokkha website for now. bdnews24.com; 29 Jan 2021.

[38] World Health Organization. COVID-19 Bangladesh Dashboard. <https://covid19.who.int/region/searo/country/bd>; 2022 [accessed 27.06.22].

[39] CovidVax Live, CovidVax Live Tracker, Bangladesh, <https://covidvax.live/location/bgd>; 2022 [accessed 27.06.22].

[40] Social Security Policy Support (SSPS) Programme. rethinking social protection responses to the COVID-19 crisis: issues and policy priorities for Bangladesh. Dhaka: Government of the People's Republic of Bangladesh; 2022.

[41] Rahman D.T. Cash support for the poor in informal sector during COVID-19 pandemic. The Business Standard; 7 Jun 2020.

[42] Afsana K, Wahid SS. Health care for poor people in the urban slums of Bangladesh. Lancet 2013;382(9910):2049−51. Available from: https://doi.org/10.1016/S0140-6736(13)62295-3.

[43] Mberu BU, Haregu TN, Kyobutungi C, Ezeh AC. Health and health-related indicators in slum, rural, and urban communities: a comparative analysis. Global Health Action 2016;9:33163. Available from: https://doi.org/10.3402/gha.v9.33163.

[44] ElFeky S. Community health workers: a strategy to ensure access to primary health care services. Cairo: World Health Organization Regional Office for the Eastern Mediterranean; 2016.

[45] Mahmud N, Rodriguez J, Nesbit J. A text message-based intervention to bridge the health care communication gap in the rural developing world. Technology and Health Care 2010;18 (2):137−44. Available from: https://doi.org/10.3233/THC-2010-0576 PMID: 20495253.

[46] Curioso W.H., Karras B.T., Campos P.E., Buendía C., Holmes K.K., Kimball A.M. Design and implementation of Cell-PREVEN: a real-time surveillance system for adverse events using cell phones in Peru. AMIA Annual Symposium Proceedings Archive; 2005:176-80.

[47] Eng E, Young R. Lay health advisors as community change agents. Journal of Family and Community Health 1992;15(1):24−40.

[48] Bressler M, Lingafelter T. Neighbor helping neighbor for health care: rural community health advisor program. In: Troxel JP, editor. Government works: profiles of people making a difference. Alexandria: Miles River; 1995.

[49] Wieser C., Bruhn M., Kinzinger J., Ruckteschler C., Heitmann S. The impact of mobile money on poor rural households: experimental evidence from Uganda. Policy Research Working Paper; No. 8913. Washington, DC: World Bank; 2019.

[50] Batista C., Vicente C.P. Is mobile changing rural africa: evidence from a field experiment. NOVAFRICA Working Paper Series wp1805. Lisbon: Universidade Nova de Lisboa; 2021.

[51] Mel S.D., McIntosh C., Sheth K., Woodruff C. Can mobile-linked bank accounts bolster savings? Evidence from a randomized controlled trial in Sri Lanka. NBER Working Papers 25354. Cambridge (MA): National Bureau of Economic Research; 2018.

[52] Aggarwal S., Brailovskaya V., Robinson J. Cashing in (and out): experimental evidence on the effects of mobile money in Malawi. AEA Papers and Proceedings. 2020;110:599-604. doi: 10.1257/pandp.20201087.

[53] Jack W, Suri T. Risk sharing and transactions costs: evidence from Kenya's mobile money revolution. The American Economic Review 2014;104(1):183−223.

[54] BRAC. BRAC, partner NGOs to mobilise communities in 38 districts under high covid risk, 58 million people to reap the benefit, <https://www.brac.net/latest-news/item/1306-brac-partner-ngos-to-mobilise-communities-in-38-districts-under-high-covid-risk-58-million-people-to-reap-the-benefit>; 2021 [accessed 27.06.22].

[55] BRAC James P Grant School of Public Health. 2022 Journey towards mobile financial service: The BRAC Community Health Worker Experience, Factsheet. Dhaka: BRAC James P Grant School of Public Health, <https://bracjpgsph.org/assets/pdf/research/research-reports/Journey%20Towards%20Mobile%20Financial%20Service_%20The%20BRAC%20Community%20Health%20Worker%20Experience.pdf>; 2022 [accessed 27.08.23].

Section IV

Vulnerable populations and pandemic response

Defining vulnerable populations

People with lower socioeconomic status (SES) faced elevated risks of both contracting and dying from COVID-19. Indeed, these health disparities were closely tied to race, ethnicity, and socioeconomic class—particularly among populations such as the housing insecure; residents of overcrowded housing complexes, slums, and prisons; forcibly displaced people; migrant workers; people with disabilities and underlying medical conditions; and people who face discrimination due to class, religion, sexuality, race, ethnicity or other identities [1−4]. Low SES populations in cities often live in areas with greater pollution, poor water and sanitation, crowding, and substandard housing [5−7]. Public health interventions designed to reach these vulnerable groups were instrumental to effective outbreak responses, particularly because their unique risk factors often amplified the already elevated risk of contracting COVID-19 in densely populated urban environments.

What is unique about vulnerable populations in urban environments?

Infectious disease spread among vulnerable populations in cities constitutes both a humanitarian and a public health emergency. Even before COVID-19, discrimination and lower SES were strongly linked to higher overall rates of morbidity and mortality [8,9]. These inequities originated, in part, due to centuries of structural racism that limited access to equitable care for certain racial groups in the United States and South Africa, and across castes in India [8−11]. During the pandemic, disproportionately high rates of comorbidities paired with chronic stress further increased vulnerable populations' risk of infection and severe complications [12,13]. Further, seroprevalence studies indicated that urban slums and overcrowded housing settlements were hotspots for COVID-19 transmission. It is, therefore, instrumental to build strategies that facilitate social distancing in housing developments with high population density, improve sanitation and hygiene, and expand access to testing and care to reduce morbidity and mortality from infectious disease, while also stemming overall disease spread.

What challenges did vulnerable populations face during COVID-19?

Vulnerable populations were often unable to comply with nonpharmaceutical interventions designed to mitigate disease spread in the absence of vaccines and effective therapies, especially in urban environments. People living in crowded conditions could not socially distance or isolate themselves from their households [14,15]. People of lower SES, particularly migrants, were more likely to be employed in the informal sector or deemed essential workers [4,16]. Complying with lockdowns or distancing recommendations risked their employment status, exacerbating housing and food insecurity. Millions of people living in urban slum areas with poor water infrastructure would often leave their homes multiple times per day to acquire water and/or food, making everything from handwashing to quarantining a challenge [6,17]. If they became sick due to these challenges, racial and ethnic minorities were also more likely to face discrimination in quality and access to care. COVID-19-related closures also interrupted essential health, nutrition, and social services and produced financial stressors that reduced people's ability to afford health care [18]. For example, as previously discussed, this phenomenon challenged public health officials in Idlib, Syria, where individuals living in camps for displaced people were unable to socially distance themselves because of high population density and crowding in the camps, and they needed to queue for food and water.

What strategies were used by governments to address challenges faced by vulnerable populations?

Several strategies were available to governments, public health institutions, and the private sector to reduce transmission among these vulnerable groups [6]. Major urban settings in the United States, such as New York City, Chicago, Baltimore, Los Angeles, Seattle, and Miami offered infected, exposed, and/or homeless individuals access to free hotel rooms to prevent case clusters in households and informal housing settings [19]. Infection prevention and control measures including the provision of personal protective equipment, engineering controls (e.g., physical barriers, ventilation improvements), enhanced disinfection procedures, testing, and vaccination helped limit transmission in workplaces delivering essential services [20]. The following chapters outline how home deliveries of food and water to homes helped reduce the number of trips individuals had to make in Mumbai's Dharavi slum, and New Orleans used easily accessible "hyperlocal" testing centers to support populations unable to use drive-through centers.

Chapter introductions

The following section contains chapters that outline experiences in pandemic response in several cities—Cape Town, South Africa; Dharavi Slum, Mumbai, India; and New Orleans, United States of America—and how health authorities and workers addressed the unique challenges faced by vulnerable populations in these respective cities. These chapters do well to illustrate how modifying different aspects of a pandemic response can protect both the most vulnerable populations in a city, as well as the local population as a whole.

The City of New Orleans has a unique geography and centuries of history guiding its modern disaster response (Fig. 1). Building on experiences in responding to infectious outbreaks accrued over nearly 200 years, the city was able to not only respond but also quickly adapt to the enormous and rapidly evolving needs that characterized the early phases of the COVID-19 pandemic. Relying upon collaboration and an intimate knowledge of local contexts to guide efforts, authorities in the city were able to direct resources to better promote an equitable response and meet the needs of some of the city's most vulnerable residents who

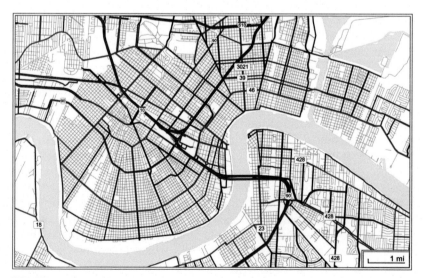

FIGURE 1 A map of New Orleans, Louisiana, United States of America. Base map sourced from OpenStreetMap, using materials made available under CC BY-SA 2.0.

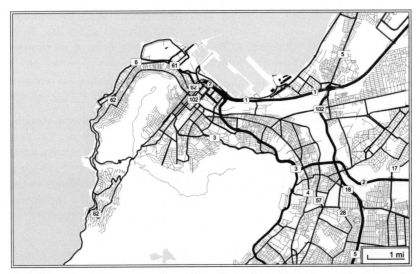

FIGURE 2 A map of Cape Town, Western Cape, South Africa. Base map sourced from OpenStreetMap, using materials made available under CC BY-SA 2.0.

otherwise may have been overlooked and systemically disenfranchised by the outbreak due to historical legacies.

The chapter on Cape Town, South Africa discusses how authorities employed a "hot-spot" strategy to focus resources on responding to low-income areas with a high burden of disease (Fig. 2). The chapter also discusses how the pandemic response efforts led to feelings of confusion and despair, and how, in addition to the health components, pandemic response efforts must also work to restore dignity and provide a sense of hope.

Dharavi, a slum in Mumbai, India is one of the largest and most densely populated slums in the world (Fig. 3). Because of this, when the first case of COVID-19 was reported on April 1, 2020, certain response measures, such as practicing social distancing, were impractical at best, and impossible at worst. Despite these challenges, it was imperative to mount a response to the outbreak and in the context of rapidly increasing case counts, active screening of all slum residents was pursued, and contact tracing of confirmed cases was conducted. Those that were symptomatic were isolated at local sports clubs and schools that had been converted to temporary health facilities or treated at a private hospital that had been acquired by authorities. Local authorities also took action to ensure that essential goods were made available to all slum residents and that their cultural norms were respected to the greatest extent possible.

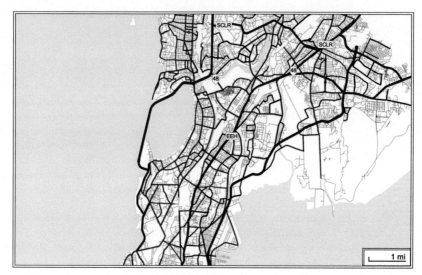

FIGURE 3   A map of Mumbai, Maharashtra, India. Base map sourced from OpenStreetMap, using materials made available under CC BY-SA 2.0.

References

[1]   Politi J, Martín-Sánchez M, Mercuriali L, Borras-Bermejo B, Lopez-Contreras J, Vilella A, et al. Epidemiological characteristics and outcomes of COVID-19 cases: mortality inequalities by socio-economic status, Barcelona, Spain, 24 February to 4 May 2020. Eurosurveillance 2021;26(20):2001138. Available from: https://doi.org/10.2807/1560-7917. ES.2021.26.20.2001138.

[2]   Hawkins RB, Charles EJ, Mehaffey JH. Socio-economic status and COVID-19−related cases and fatalities. Public Health 2020;189:129−34. Available from: https://doi.org/ 10.1016/j.puhe.2020.09.016.

[3]   Calderón-Larrañaga A, Vetrano DL, Rizzuto D, Bellander T, Fratiglioni L, Dekhtyar S. High excess mortality in areas with young and socially vulnerable populations during the COVID-19 outbreak in Stockholm Region, Sweden. BMJ Global Health 2020;5(10): e003595. Available from: https://doi.org/10.1136/bmjgh-2020-003595.

[4]   World Health Organization. Regional Office for the Western Pacific. Actions for consideration in the care and protection of vulnerable population groups for COVID-19 Report No.: WPR/DSE/2020/021. Manila: WHO Regional Office for the Western Pacific; 2020.

[5]   Williams DR. Race, socioeconomic status, and health the added effects of racism and discrimination. Annals of the New York Academy of Sciences 1999;896(1):173−88. Available from: https://doi.org/10.1111/j.1749-6632.1999.tb08114.x.

[6]   Auerbach AM, Thachil T. How does Covid-19 affect urban slums? Evidence from settlement leaders in India. World Development 2021;140:105304. Available from: https://doi. org/10.1016/j.worlddev.2020.105304.

[7]   Evans GW, Kantrowitz E. Socioeconomic status and health: the potential role of environmental risk exposure. Annual Review of Public Health 2002;23(1):303−31. Available from: https://doi.org/10.1146/annurev.publhealth.23.112001.112349.

[8]   Hogarth RA. Medicalizing blackness: making racial difference in the Atlantic. World. Chapel Hill: University of North Carolina Press; 2017. p. 1780−840.

[9]   Washington HA. Medical apartheid: the dark history of medical experimentation on Black Americans from colonial times to the present. New York: Doubleday Publishing Company; 2006.

[10] Mooney GH, McIntyre DE. South Africa: a 21st century apartheid in health and health care. Medical Journal of Australia 2008;189(11):637−40. Available from: https://doi.org/10.5694/j.1326-5377.2008.tb02224.x.

[11] Mahapatro SR, James KS, Mishra US. Intersection of class, caste, gender and unmet health care needs in India: Implications for health policy. Health Policy OPEN 2021;2:100040. Available from: https://doi.org/10.1016/j.hpopen.2021.100040.

[12] Segerstrom SC, Miller GE. Psychological stress and the human immune system: a meta-analytic study of 30 years of inquiry. Psychological Bulletin 2004;130(4):601−30. Available from: https://doi.org/10.1037/0033-2909.130.4.601.

[13] Wiemers EE, Abrahams S, AlFakhri M, Hotz VJ, Schoeni RF, Seltzer JA. Disparities in vulnerability to complications from COVID-19 arising from disparities in preexisting conditions in the United States. Research in Social Stratification and Mobility 2020;69:100553. Available from: https://doi.org/10.1016/j.rssm.2020.100553.

[14] Shaw JA, Meiring M, Cummins T, Chegou NN, Claassen C, Plessis ND, et al. Higher SARS-CoV-2 seroprevalence in workers with lower socioeconomic status in Cape Town, South Africa. PLoS One 2021;16(2):e0247852. Available from: https://doi.org/10.1371/journal.pone.0247852.

[15] Surendra H, Salama N, Lestari KD, Adrian V, Widyastuti W, Oktavia D, et al. Pandemic inequity in a megacity: a multilevel analysis of individual, community and health care vulnerability risks for COVID-19 mortality in Jakarta, Indonesia. BMJ Global Health 2022;7(6):e008329. Available from: https://doi.org/10.1136/bmjgh-2021-008329.

[16] The Lancet. The plight of essential workers during the COVID-19 pandemic. Lancet 2020;395(10237):1587. Available from: https://doi.org/10.1016/S0140-6736(20)31200-9.

[17] Staddon C, Everard M, Mytton J, Octavianti T, Powell W, Quinn N, et al. Water insecurity compounds the global coronavirus crisis. Water International 2020;45(5):416−22. Available from: https://doi.org/10.1080/02508060.2020.1769345.

[18] Pereira Bajard M, Stephens N, Eidman J, Warren KT, Molinaro P, McDonough-Thayer C, et al. Serving the vulnerable: The World Health Organization's scaled support to countries during the first year of the COVID-19 pandemic. Frontiers in Public Health 2022;10:837504. Available from: https://doi.org/10.3389/fpubh.2022.837504.

[19] Allison TA, Oh A, Harrison KL. Extreme vulnerability of home care workers during the COVID-19 pandemic—a call to action. JAMA Internal Medicine 2020;180(11):1459−60. Available from: https://doi.org/10.1001/jamainternmed.2020.3937.

[20] Carlsten C, Gulati M, Hines S, Rose C, Scott K, Tarlo SM, et al. COVID-19 as an occupational disease. American Journal of Industrial Medicine 2021;64(4):227−37. Available from: https://doi.org/10.1002/ajim.23222.

Chapter 9

Prioritizing local context and expertise in a global pandemic: the New Orleans response to COVID-19

Jennifer Avegno[1,2], Kasha Bornstein[2] and Jordan Vaughn[2]
[1]New Orleans Health Department, New Orleans, LA, United States, [2]Department of Emergency Medicine, LSU Health Sciences Center New Orleans, New Orleans, LA, United States

Background

Historical context of the city of New Orleans

The city of New Orleans, in the state of Louisiana, occupies a unique place in geography and history, making it repeatedly resilient in the face of a variety of health and life threats. One of the largest "majority-minority" cities in the United States, with a population that is 59% Black, 5.5% Hispanic or Latinx, and 3% Asian [1], New Orleans' history is deeply rooted in a multicultural and syncretic coalescence of communities with distinct trajectories. Inhabited primarily by the Chitimacha tribe for thousands of years during the precolonial period, the area that would become New Orleans was first colonized by the French in 1718. New Orleans was the territorial capital of French Louisiana, saw rule by the Spanish Empire in the latter half of the 18th century, and was briefly under French control again before it was sold to the United States as part of the Louisiana Purchase in 1803 to finance unsuccessful efforts to reassert French rule over Haiti following the Haitian revolution.

During the 19th century, New Orleans became the most productive port in the United States, the largest city in the South, the third most populous city in the United States, and between 1820 and 1920, the leading point of entry for immigration after New York City [2]. As a central port city in the antebellum south, New Orleans also functioned as the center of both the slave trade and products produced by enslaved peoples, including cotton and sugar. Hundreds of thousands of enslaved African people were trafficked through the city's port and markets, many of whom would remain in southern Louisiana to labor on the plantations in the region. As a commercial and cultural center, New Orleans was also home to one of the most significant populations of free people of color, particularly as Francophone Haitians moved to the city in the period following the Haitian revolution.

New Orleans remained an economically, culturally, and socially diverse city after the US Civil War and into the 20th century. Waves of immigrants from Ireland, Germany, and Italy throughout the latter period of the 1800s intermixed European cultural elements with local traditions and brought greater complexity to rapidly changing racial politics of governance throughout the Reconstruction period. During the post-war period, New Orleans saw Black legislators elected to local and state offices, but also a retrenchment of racial animus in numerous race riots and efforts to restore *de jure* white supremacy in city and state governance.

The long history of free people of color in the city, large populations of educated Black residents, and mixed-race families—many of whom self-describe as Creoles—sets New Orleans aside as a relatively liberal city in the Deep South. It was a center of the Civil Rights movement throughout the 20th century with an early statewide push for progressive racial policy under the tenure of Huey P. Long, and a multidecade clash with the state government in determining educational, economic, and public health policy. Over the second half of the 20th century, like much of the United States, New Orleans saw white flight, redlining, resegregation, and structural disinvestment as middle-class post-war white families left for regional suburbs. These demographic shifts deeply affected the sociopolitical circumstances of the city, now a predominantly Black, progressive enclave in an otherwise deeply conservative southern state [3].

Federal, state, and local governmental disinvestment and isolation precipitated many of the structural failures undergirding the dysfunctional regional preparation for Hurricane Katrina in 2005, as well as the economic, ecological, public health, and social sequela that followed that disaster. Over 1 million people fled the central Gulf Coast, becoming the largest diaspora in the United States since the Great Migration of southern Black families in the early 20th century [4]. The city of New Orleans saw over half of its population leave in the immediate post-Katrina period, and as of 2020 had only recovered approximately 80% of the pre-Katrina count of 484,674 [4]. Nonetheless, this period saw an amplification in immigration from Mexico and Central America, as workers moved to the region to help rebuild the city.

Today, communities in New Orleans continue to navigate the systemic and social effects of this complex history. The poverty rate is approximately 25%, and the structural effects of poverty and high social vulnerability play out biologically: the average life expectancy in the New Orleans metro is 76.6, lower than the national average and among the lowest of all metro areas across the United States [5].

Public health in New Orleans

The COVID-19 pandemic is far from the first major infectious disease outbreak confronted by New Orleans. The city faced numerous eruptions of yellow fever throughout the 19th century, including the 1853 outbreak that killed 7849 and sickened almost 30,000 [6]; at least 41,000 people died of the disease between 1817 and 1905 [6]. Prior to the identification of *Aedes aegypti* as the primary vector for yellow fever, it was popularly believed to derive from miasma—humid air acting on filthy, undrained soil. Although the local Department of Public Works was responsible for maintaining drainage, the city's topography below sea level and its proximity to both Lake Pontchartrain and the Mississippi River meant that dirty water stagnated and flooded regularly; employees cleaned gutters and canals by shoveling muck onto the streets, only to have it wash back during the next rain (Fig. 9.1).

Blaming infectious disease outbreaks on poor residents and immigrants, businessmen and politicians who were invested in the continued flow of goods and people into Louisiana ignored or purposefully covered up disease and death. To maintain an open port free of quarantines, business interests tried to convince newspapers and directories not to publish negative news or publicize the astounding number of deaths in New Orleans [6]. Local politicians who controlled New Orleans during much of the period were reluctant to spend money on preventing or stopping the spread of pestilence as most of those dying of yellow fever were most often Black or Irish, German, and French immigrants [7].

During the peak of the rainy season during epidemic periods of yellow fever, wealthy white families would flee the summer heat by retreating upriver, often spreading infection along their route. In 1878 a devastating yellow fever outbreak originating in New Orleans spread as far as Memphis, Tennessee, killing almost 20,000 people across the Mississippi River valley. Upriver towns and neighboring Gulf Coast cities like Mobile, Alabama shut down all travel to and from New Orleans, closing trade with the city at the first hint of disease. When yellow fever struck again in 1897, it killed almost 300 people, the highest number of deaths since 1878. Officials and the population alike were frightened into action, and in 1899 voters, encouraged by a group of Progressive reformers known as the Municipal Improvement Association, at last, approved funds for a drainage and sewerage system that would permanently clean up the city's stagnant waters.

It was in large part due to the city's repeated exposure to yellow fever outbreaks that the New Orleans Health Department (NOHD) was created. As early as 1854, Dr. Edward H. Barton, a prominent local physician and nascent public health practitioner, recommended in an official report of the 1853 yellow fever outbreak that a Department of Health be established in New Orleans [8]. Though some functions to "maintain cleanliness and health" existed in municipal law thereafter, state law officially established a

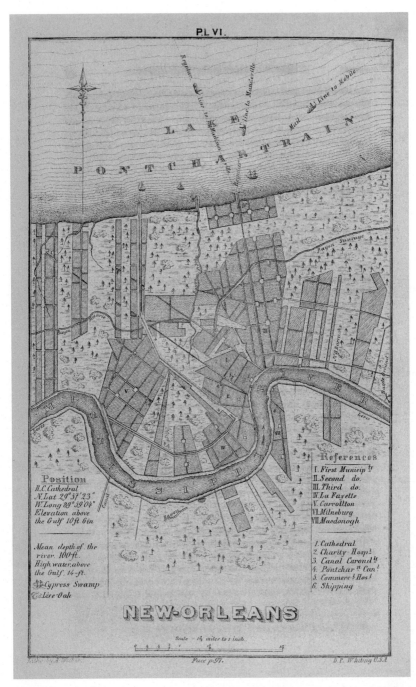

FIGURE 9.1 1850 Map of the City of New Orleans. *Daniel Drake, Early Printed Books, a systematic treatise, historical, etiological, and practical, on the principal diseases of the interior valley of North America: as they appear in the Caucasian, African, Indian, and Esquimaux varieties of its population. Available under Creative Commons Attribution Only License CC BY 4.0.*

Board of Health for the city of New Orleans in 1898, and formally transitioned to a Department of Health—separate from but accountable to the State Health Department—in the mid-20th century [9]. Since its early days, the Board or Department of Health had major responsibilities to investigate and treat infectious disease outbreaks and epidemics within the city.

Presently, the NOHD is part of the city's Executive Branch and is led by a Director of Health appointed by the Chief Administrative Officer with mayoral approval [10]. The Director is expected to have significant experience in the administration of health programs in either the public or private sector or to possess an advanced degree in medicine, nursing, or public health [10].

In New Orleans, the Health Department is responsible for protecting the health of its city by enforcing the State sanitary code, enforcing and adopting local health and sanitary regulations, investigating the causes and instituting measures for the control of epidemic, preventable communicable diseases, and operating emergency medical services, subject to applicable state and municipal law [10]. As the relationship between state and local health departments has evolved over the past century, NOHD generally provides enhanced, supportive, or nonduplicative services to Louisiana Department of Health (LDH) functions. Through formal and informal agreements, NOHD and LDH partner on larger state and regional public health threats or responses (e.g., epidemics, disasters, overall community health improvement plans), but NOHD has many independent programs and initiatives solely focused on the residents of Orleans Parish. In cases of "infectious, contagious, or pestilential disease" the Director of Health has specific and independent powers to enact quarantine regulations and control "access to or egress from" any building or place if it may be necessary to prevent the spread of disease [11]. This authority is not dependent on emergency orders or declarations by the State, though the Governor or Mayor could issue executive orders or proclamations in a state of disaster or emergency with other specific regulations that may be enforced by the NOHD. However, constitutional limits on state authority and any guaranteed rights remain in full effect during an emergency [12]. These independent powers of the NOHD in times of infectious disease outbreaks stem from previous epidemics and were designed to protect the public's health against other societal and economic forces that may be in conflict.

Previous public health emergencies in New Orleans

New Orleans is no stranger to public health disaster response. For example, the city responded to the influenza pandemic of 1918, known as Spanish Influenza, in a manner particular to its history and culture. When Spanish Influenza struck the city between September 1918 and April 1919, some 54,089 were sickened and 3489 people died—nearly 1% of the city's population, about twice the national rate, and more than in all but the worst yellow fever outbreaks of the prior century [13]. At the time of that pandemic, New Orleans was still one of the largest port cities and a major commercial center, and the city saw the pandemic enter both by sea and by land. The influenza outbreak is thought to have begun on September 16 when an oil tanker arrived at the port of New Orleans with five crewmembers ill with influenza [14]. In contrast to the laissez-faire attitude the city had toward yellow fever over much of the 19th century, city health inspectors visited the ship and immediately quarantined the vessel. Nonetheless, cases began to appear in city hospitals, tracked to other steamships and the regional naval hospital. The State Board of Health drafted an emergency amendment to the Louisiana sanitary code, mandating physicians to report influenza cases to their local health boards, followed by the New Orleans Board of Health making influenza a mandatory reportable disease two weeks later [14].

As influenza cases soared past 7000 by October 1918, the New Orleans the mayor Superintendent of Health and the Mayor ordered to close all schools, churches, theaters, movie theaters, and prohibited public gatherings, and instructed the New Orleans Railway and Light Company to limit passengers to one passenger standing for every two passengers sitting on the city's streetcars [14]. Local civic organizations responded in kind: The motormen's union urged workers to lengthen their usual nine-hour workday and take as many extra trips as would be necessary to keep streetcars from becoming crowded [14]. Even still, cases continued to rise, challenging the capacity of local public health and medical resources. Mardi Gras festivities in 1919 were put on hold, and as New Orleans' Charity Hospital turned its facility completely over to the treatment of influenza patients, a coalition of city, state, national, and non-governmental services directed resources toward building emergency hospitals and financing staffing. More than 4000 city government employees and factory workers assisted with voluntary vaccination clinics, while medical students were deputized as assistant US Public Health Service surgeons to provide care in newly established influenza wards [14]. Cases began to abate early into 1920, the result of local, state, and federal coordination of aid, services, and implementation of public health measures.

The early 21st century has also provided the city with multiple modern opportunities to design, test, and implement response plans for large-scale outbreaks. In 2007 the LDH, partnering with the NOHD and others, scheduled a training exercise to help New Orleans and its surrounding areas prepare for a future influenza pandemic [15]. The training consisted of nine regional exercises that featured discussion-based scenarios designed to educate each community [16]. It laid the framework to work with local, tribal, territorial, and state public health officials as they undertook critical preparedness planning. In attendance were industry leaders, public safety professionals, faith-based organizations, tribal organizations, and other federal, state, and local agencies. These exercises and simulations identified local and regional partners whose involvement would be critical in a widespread pandemic event and served to establish and solidify personal and professional relationships, as well as efficient processes and communication channels necessary for a robust response.

In 2016 during the Zika epidemic, the Mayor of New Orleans released a plan to address this emerging threat. The NOHD, the City of New Orleans Mosquito and Termite Control Board, and other partners had already planned mosquito control and public education activities to combat the mosquito-borne virus. The US Centers for Disease Control highlighted New Orleans' approach and recommended that other cities replicate their model to lead and proactively prepare for local transmission of the virus using a multiagency approach [17].

New Orleans' Zika plan was composed of multiple phases: phase one focused on education and mosquito control and took place before a single case had been diagnosed locally. The general public had access to a website with information, as well as direct field outreach to individuals, clinicians, and organizations serving pregnant people and pest control professionals [18]. Community partners assisted city personnel in getting the word out and building a knowledge base about the disease and its risks. In February 2016, after the first New Orleans area case was identified, phase two began: door-to-door campaigns were launched that focused on protection against mosquito bites, and code enforcement was increased to identify and mitigate mosquito breeding grounds. If and when a Zika outbreak expanded, phase three of the plan would activate teams to aggressively control mosquitos with aerial and ground mosquito treatments, property inspections, and more targeted educational campaigns in areas with active Zika transmission reported and confirmed [18]. The Zika epidemic proved to be an excellent real-world foundational exercise for NOHD to prepare for COVID-19 a few years later; relationships forged through community collaborations, interagency partnerships, and state and federal working groups created a "muscle memory" that was easily reawakened a few years later with the COVID-19 pandemic.

COVID-19 in New Orleans

While early 21st century outbreaks in New Orleans did not have the destructive force or impact of previous yellow fever epidemics, or of the COVID-19 pandemic to come, they did provide the groundwork for a successful emergency response. By involving multiple stakeholders and sectors, leveraging existing resources, building and refining detailed plans, and testing them through both exercises and real-world events, New Orleans entered 2020 with a solid foundation for a wider pandemic strategy.

The first confirmed COVID-19 case in New Orleans was diagnosed on Monday, March 9, 2020. This day had been anticipated for several weeks—and was not unexpected, given the rise in cases across the country at that time. Several details about the first case—and the next two that were confirmed on March 10—raised significant alarm. The Carnival season concluded almost exactly 14 days prior and comprised two weeks of huge, almost nightly, parades and large gatherings; Mardi Gras Day 2020 coincided with the first known identified case of COVID-19 from community spread in Washington State. Although most experts still assumed that domestic cases stemmed from travel to international hot spots, none of the three initial cases in New Orleans had traveled internationally or could be traced to contact with each other. In this way, it was very quickly understood that the undiagnosed outbreak was far greater than could be measured with existing tools and that the potential for danger was immense.

Multiple large gatherings and city-permitted parades were planned for the weekend of March 15; however, based on NOHD officials' understanding of community dynamics, characteristics of these first three infected citizens, and the assumed huge public risk, city administrators immediately canceled the large public weekend activities. An incident command structure was set up locally, led by the NOHD and the New Orleans Department of Homeland Security (NOHSEP), and a regular cadence of communication with the LDH, the Governor's Office of Homeland Security and Emergency Preparedness (GOHSEP), and Governor's Office was established. Because New Orleans— and Louisiana—had years of experience with crisis management of natural disasters and other infectious diseases as above, key relationships were already known and well-seasoned, and the "muscle memory" of fighting off the crisis at hand was strong.

Though the response involved experienced personnel and trusted relationships early on, during this first week of pandemic impact, the NOHD had few resources and huge knowledge gaps on the nature and transmission of the virus. Criteria to test for the presence of SARS-CoV2 remained stringent, and test availability, as well as the timeliness of results, was severely constrained in the early days of the pandemic. Within two weeks, cases had risen by a factor of 400, with a nearly 5% case fatality rate [19]. Because reliable, community-specific modeling was not available nationally, the NOHD repurposed a data unit in its local Department of Public Works to provide infection and mortality projections, as well as begin to map the outbreak across the city. Initial estimates indicated that if no lockdown or restrictive measures were implemented, New Orleans could expect to see 1300 deaths (0.3% of the current population) within eight weeks. By Friday, March 16, 2020, the city had closed schools and instituted "Stay at Home" orders—one of the first cities in the United States to do so.

The response to COVID-19 in New Orleans: reaching the most vulnerable

Because of the city's emerging status as a significant hotspot, in the second week of the local outbreak, New Orleans was approached by the federal government to design, implement, and assist in operating one of the nation's first drive-through mass testing sites. This required novel, active collaboration between federal agencies and departments (Federal Emergency Management Agency, Department of Homeland Security, Department of Health and Human Services, Public Health Service Corps); state agencies (LDH, GOHSEP, Louisiana National Guard); and local offices (NOHD, NOHSEP, as well as emergency medical services and public safety). Based on past state-and-city collaborations on outbreaks and other disasters, however, the foundation for partnerships was well-established. While the project was led and directed by federal leaders, the on-the-ground implementation and communication fell to state and local authorities (Table 9.1).

Daily calls between all partners began one week after New Orleans' first diagnosed case and continued through the initial opening of the first drive-through testing facility a few days later. As the process and plan developed, challenges with implementing a federal initiative in the context of a very local crisis came to light, and concerns were raised by state and local health representatives. As an example—federal officials insisted on National Guard personnel staffing upon entry to the site and mandated Louisiana identification (as well as specific symptoms) for anyone wishing to be tested. Federal

TABLE 9.1 Governmental agencies and contributions to New Orleans' Initial Mass Testing Sites.

Contribution	Federal	State	Local
Materials	Tests Lab analysis Site infrastructure	Site infrastructure	Site Infrastructure
Standard operating procedures/plans	Initial development	Developed and revised	Developed and revised
PERSONNEL	FEMA DHS HHS Public Health Service Corps	GOHSEP LDH Louisiana National Guard	NOHD NOHSEP EMS Regional homeland security partners Volunteers
Communications	Federal call center (test result notification); patient information	Statewide media External agencies (site selection and permissions)	Local media, flyers, patient information; External agencies (site selection and permissions)

partners also designed a call-back system for results where a remote call center would receive test information and call individuals about their test results, but not leave a message if no one answered.

Local public health personnel at NOHD are overwhelmingly longtime community residents with extensive knowledge of and experience with the city they serve. As such, they are acutely aware of the attitudes of residents—a historic mistrust of government and health personnel; transportation, language, and citizenship barriers for many; suspicion of non-native New Orleanians from outside the area inserting themselves into a disaster response with little local context (a huge issue post-Hurricane Katrina) —and how these can contribute to challenges in implementing public health interventions. As such, upon hearing the initial plan for the federal testing sights, local officials pointed out that community residents without a Louisiana ID, or with significant mistrust of the military, would be excluded from the ability to test and thus potentially worsen their health outcomes while increasing the chance of spreading the virus to others before it could be detected. Furthermore, since the call center was located several states away, little appropriate information would be available to provide anxious patients who tested positive about their next steps. State and local officials lobbied to transfer call center responsibilities to LDH and NOHD personnel and trained volunteers, who could provide a sympathetic, understanding local voice and real connection to the resources available to residents of the city. Initially, these requests were denied by Federal partners.

Three drive-through sites (Figs. 9.2 and 9.3) were ultimately opened in the New Orleans region, and within three weeks saw approximately 13,000 individuals for testing—by all accounts, a tremendous success in driving critical testing resources to the area at a time when they were still significantly limited in many other places. New Orleans rapidly went from some of the fastest growing reported COVID-19 case and death rates in the world to an outbreak that was largely contained within a month, thanks to far greater testing resources than most other cities at that time. The sites also provided a greater public good, as they not only served residents but also drew individuals from around the region and nearby states. Further, the information and experiences gained from testing processes, lab throughput times, and results were used to inform and enhance other local and national testing programs in the future.

However, throughout the sites' operating period, state and local health officials continued to advocate for a more equitable approach. Using patient address data, the NOHD mapped residences of those accessing the sites and very quickly identified some glaring gaps in citywide coverage. Not surprisingly, neighborhoods with traditionally high levels of social vulnerability and low access to transportation or health care, in general, were grossly underrepresented in the drive-through site patient population. The NOHD quickly and effectively used these findings to demonstrate to federal partners the need

FIGURE 9.2   The site of the first federally supported COVID-19 drive-through testing program in the city of New Orleans, located at the University of New Orleans Arena.

FIGURE 9.3   Typical personnel at a federal testing site—military, FEMA, HHS, and other generally non-local staff. *FEMA*, Federal Emergency Management Agency; *HHS*, Department of Health and Human Services.

for a different model of mass testing: one that would be deliberately placed at trusted community sites (churches, community centers, historically black colleges and universities [HBCUs]) in historically underresourced neighborhoods, run by local agencies and institutions and with New Orleanians providing call center duties and offering direct assistance. After much discussion, the NOHD and LDH were able to reach an understanding with federal partners to be able to utilize some of their key resources (i.e., tests and laboratory analysis) in this new, equity-focused, and hyperlocal community test site model.

The first community-based test site opened in mid-April 2020 in the parking lot of a centrally located, well-respected HBCU in a neighborhood comprised mainly of residents of color (Figs. 9.4 and 9.5). It was a success—lines stretched around the block an hour prior to opening time, confirming the unmet demand and need for better access to testing services. Over 50 unique subsequent sites were deliberately placed in areas with partners focused on groups disproportionately affected by COVID-19 due to pre-existing inequities. For example, the NOHD and LDH designed a patient intake system that mandated ethnicity as part of demographic information gathering, and set up an early site, with multiple fluent non-English speakers and resources in Spanish, at a church that maintained a well-established and trusted relationship with Spanish-speaking communities. This allowed early identification of extremely high rates of infection in the New Orleans' Hispanic communities and their unique vulnerabilities to a worsened outbreak. Because of the NOHD's continued contact with federal partners, these findings could be communicated directly, and these local efforts were recognized by White House COVID-19 response leaders as sounding the early national alarm for the disproportionate impacts of the virus on Hispanic communities.

FIGURE 9.4 New Orleans' first community test site at Xavier University, a centrally located and well-respected historically black university.

FIGURE 9.5 New Orleans Health Department staff and local partners operating an early community test site in a large, easily accessible neighborhood of color.

Further evidence of the success of this model shift came from a research study on the effectiveness of community sites on access to testing services for at-risk populations [20]. Comparing the original federally directed drive-through sites with the community testing model, researchers found that neighborhood locations increased testing availability, particularly in neighborhoods of color, for elderly patients, and among Black and Asian residents [20]. Data from these testing sites showed that Hispanics were more likely to travel long distances to access the community sites, possibly because attention was made to ensuring an inclusive, welcoming, low-barrier environment.

Nationally, these types of community testing sites that were championed by New Orleans soon became the norm for increasing access, and public-private-academic partnerships similar to this model blossomed. And, in New Orleans, the NOHD was able to replicate and refine this process for vaccination sites once vaccines were available later in 2020 and 2021. The NOHD once again collaborated on a mass vaccination site with federal partners for funding and resources, the state and local health care system for management and personnel, and tens of thousands of residents were vaccinated there— with fewer barriers, transportation assistance, and a more inclusive environment. However, the bulk of the local on-the-ground efforts continued to revolve around businesses and trusted spaces in communities of color— churches, bars, small businesses, and clubs—with culturally competent ambassadors and health care personnel to answer questions and reduce mis- and disinformation. The NOHD strongly believes that the commitment to equity from the beginning of testing helped increase buy-in from historically hesitant populations, resulting in high vaccination rates for New Orleans, and the state of Louisiana having higher vaccination rates in Black populations when compared to white populations.

Conclusion

While gaps and disparities remain, the overall foundation of bringing all partners to the table—federal, state, municipal, private, and nonprofit—to quickly and flexibly design a system focused on equity and local context continues. Strengthening these personal relationships and formal channels of communication and division of labor will continue to serve the NOHD and the city well through the duration of the COVID-19 pandemic and in other future crises in New Orleans. The weight and experience of history cannot be understated in a modern response to crisis. Truly, the terrible lessons learned from structural racism amidst yellow fever, influenza, and other disasters faced in past centuries must be considered and actively adapted to protect the community far better in this era.

References

[1] United States Census Bureau, 2020. Racial and ethnic statistics, <https://data.census.gov/cedsci/table?g = 1600000US2255000&tid = DECENNIALPL2020.P2>; 2020 [accessed 22.08.22].

[2] LeMay MC. Doctors at the borders: immigration and the rise of public health. Westport: Praeger; 2015.

[3] Fussell E. Constructing New Orleans, constructing race: a population history of New Orleans. Journal of American History 2007;94(3):846−55. Available from: https://doi.org/10.2307/25095147.

[4] Rhodes J, Chan C, Paxson C, Rouse CE, Waters M, Fussell E. The impact of Hurricane Katrina on the mental and physical health of low-income parents in New Orleans. American Journal of Orthopsychiatry 2010;80(2):237−47. Available from: https://doi.org/10.1111/j.1939-0025.2010.01027.x.

[5] Habans R., Losh J., Weinstein R., Teller A. Placing prosperity: neighborhoods and life expectancy in the New Orleans Metro, <https://www.datacenterresearch.org/placing-prosperity>; 2020 [accessed 22.08.22].

[6] Drakeford S. Resource set: yellow fever. New Orleans: The Historic New Orleans Collection; 2020.

[7] Perkins E, Magill J. In the late 1800s, devastating yellow fever epidemics forced New Orleans to confront its sanitation problem. New Orleans: The Historic New Orleans Collection 2020.

[8] Postell WD. Edward Hall Barton, Sanitarian. Annals of Medical History 1942;4(5):370−81.

[9] New Orleans Health Department. Historical note, City of New Orleans Archives, <http://archives.nolalibrary.org/~nopl/inv/health.htm>; 2001 [accessed 22.08.22].

[10] Council of the City of New Orleans. Home rule charter of the City of New Orleans. New Orleans: Council of the City of New Orleans; 2010.

[11] Council of the City of New Orleans. Code of ordinances, New Orleans; Chapter 82: health and sanitation. New Orleans: Council of the City of New Orleans.

[12] State of Louisiana. Louisiana Revised Statutes §724. Powers of the Governor. Baton Rouge: Louisiana State Legislature; 2021.

[13] Board of Health for the Parish of Orleans and the City of New Orleans. Biennial report, 1918−1919. New Orleans: Brandao Printing Co.; 1919.

[14] University of Michigan Center for the History of Medicine. New Orleans, Louisiana and the 1918−1919 influenza epidemic. Ann Arbor: Michigan Publishing; 2016.

[15] U.S. Centers for Disease Control and Prevention. 2009 H1N1 pandemic (H1N1PDM09 virus), <https://www.cdc.gov/flu/pandemic-resources/2009-h1n1-pandemic.html>; 2010 [accessed 22.08.22].

[16] Louisiana Department of Health. Pandemic flu response training set for New Orleans Area. Baton Rouge: LDH; 2007.

[17] U.S. Centers for Disease Control and Prevention. Proactivity and partnerships: New Orleans prepares for zika. Atlanta: CDC; 2016.

[18] City of New Orleans. Condensed Zika virus plan. New Orleans: City of New Orleans; 2016.

[19] United States Centers for Disease Control and Prevention. CDC covid data tracker, <https://covid.cdc.gov/covid-data-tracker/>; 2022 [accessed 23.08.22].

[20] Hernandez JH, Karletsos D, Avegno J, Reed CH. Is Covid-19 community level testing effective in reaching at-risk populations? Evidence from spatial analysis of New Orleans patient data at walk-up sites. BMC Public Health 2021;21(1). Available from: https://doi. org/10.1186/s12889-021-10717-9.

Chapter 10

COVID-19, control and confusion: experiences from Cape Town, South Africa

Patricia van der Ross
Community Services and Health, City of Cape Town, South Africa

Background

The city of Cape Town, located in the Western Cape Province of South Africa, is a city known for providing all of the benefits of city living right at the beachside, with a view of one of the Seven Wonders of the World, Table Mountain. The city and the country have come a long way since the abolishment of Apartheid in 1994. In South Africa, the Apartheid government was a regime known for its unequal treatment of human beings, based on the color of their skin—with those with light skin tones enjoying greater rights and privileges than those with darker tones. However, in 1994, a new constitution that enfranchised Black citizens and other racial groups took effect, marking the official end of the apartheid system and ushering in a new era of democratic governance. In comparison to other developing countries, South Africa is considered a relatively young democracy. Many have argued that inequalities remain deep, some arguing that this should be attributed to the lack of political will and economic growth.

While the social-political economy may be seen as a weakness given the country's history and incremental change, its greatest strength lies in the diversity of its people, which is sourced from rich cultures and traditions. This vibrancy and diversity draw tourists to the country—providing many benefits, especially those of income and prosperity.

With a population of roughly 4.8 million people, the city of Cape Town in Western Cape is the second most populous city in South Africa, second only to the city of Johannesburg. As of 2020, according to the Western Cape Government, the city is, broadly speaking, demographically divided by both economics and language. The ethnic and racial composition is roughly 80%

Black and Colored, with White, Indian, and "Others" constituting the remaining 20%.

The city is also rapidly growing, and recent demographic trends include an annual growth rate of 2% [1]. Many have argued that the city may be unable to sustain this rapid population growth, and statistics recently revealed by the Minster of Local Government in the Western Cape indicated a need for an entirely new city in a matter of eight years given the current population growth rate [2].[1] Further, because of this rapid growth, Cape Town is a city that is all too familiar with economic strife and societal inequality. Some of this rapid population growth may be attributed to the availability of employment opportunities and improved livelihoods compared to other locations in South Africa [3], and as a result, the city hosts a great deal of wealth. However, because of persistent inequalities, at the other end of this spectrum are individuals who are entirely dependent on the social security nets provided by national and subnational governments.

The public health system in Cape Town

The South African state is governed by three distinct spheres of government—the National Government, nine provincial governments, and local governments (i.e., approximately 280 municipalities, countrywide). The National Government sets policies for the country, which are then executed and implemented nationwide by the provincial and local governments. Provincial and local governments are thus the levels of government that are primarily responsible for executing functions in designated jurisdictions, as mandated via Acts of Government. The provincial governments oversee the local governments, which is the sphere that is the closest to the residents.[2]

The government closest to the people is thus tasked with meeting the demands for delivering everyday public goods and services, while the higher levels of government are tasked with determining broader policy direction and necessary functions attached thereto. Altogether, the three spheres aim to separate powers and improve accountability for the delivery of public services.

The Constitution of the Republic of South Africa provides the right to health care to all South Africans, and the National Health Act 61 of 2003 provides a framework for a structured uniform health system that accounts for the obligations imposed by the Constitution and other laws passed by the national, provincial, and local governments with regard to health services [4].

---

1. Common sense tells us that there is simply no way, a similar sized city could be built and supplied with the necessary resources in this amount of time.
2. The city of Cape Town represents one example of a local Metropolitan Municipal government.

As a whole, South Africa may have one of the best health systems across the African continent, as the quality of the services provided may be measured through life expectancy, which represents one key measure for assessing health systems. Through a system of some 422 hospitals and 3,941 health clinics, the National South African Health Department administers the bulk of South Africa's health care service plans [5]. Policy and statutory guidelines for the allocation of health expenditures show that 86% of the national health budget is transferred to the provincial governments, which are financially empowered to execute health functions at the direction of the National Government. The provincial government then offers primary (i.e., every day, out-of-hospital medical care), secondary (i.e., in-hospital medical care), and tertiary health care services to the population.

Health care is thus a "big-ticket" budget item in South Africa. Still, since the end of the Apartheid, the National Government has been led by one political party that has arguably deprioritized health and health care since the start of its administration. The National Government is also known for general macro-policy abandonment, institutional inertia, and a lack of accountable leadership. Further, the national government has recently moved forward with budget cuts to the health sector in arguments of pursuing a better approach at a later date. More specifically, these cuts to the health sector are being made to make room for the national health insurance scheme that will provide universal health coverage to the entire country—all while it continues to battle with slow change and growth of infrastructure, and years of unfulfilled commitments [6].[3]

This has, unsurprisingly, resulted in a tremendous strain on the capacities required to provide even the most basic of health services in the city of Cape Town. The Western Cape is known for spending the most on public health, in terms of spending per capita, when compared to other South African provinces—with a total of approximately R3,800 per person per year, and most of this funding going toward health care workforce costs [5]. Because of this deliberate prioritization of health, the Western Cape Province has the highest life expectancy of all South African provinces.

The crux, however, lies therein. Health services are administered to the population through two streams: public health care and private health care. While roughly 80% of the population is reliant on public health care, the remaining 20%, which constitutes the wealthiest portion of the population, relies on private health care. The health services provided through the private sector are widely recognized to be of better quality, when compared to those offered through the public health care system. It is also widely recognized that the majority of specialized health services are provided by the private sector. This highlights a clear imbalance in the overall structure of the health

---

3. Universal health coverage refers to the ability to access health care when and where it is needed without financial hardship.

system and the quality of health services across the country—underscoring the importance of restoring the dignified delivery of public services that have been lost for some time.

We have been aware of the cultural political economy when it comes to the health structure in South Africa, but one thing is for sure. COVID-19 laid bare the awareness thereof. It exposed wide gaps that exist within society and wealth—essentially health.

COVID-19 in Cape Town, South Africa

Cape Town has a history of responding to infectious disease events—most notably the HIV/AIDS outbreak that began in South Africa around 1982 [7]. Still, considering this, as well as other notable communicable diseases like tuberculosis, South Africa had not faced a public health crisis on the magnitude of the COVID-19 pandemic. The emergence of COVID-19, its high degree of transmissibility and the far-reaching societal impacts mean that even while considering South Africa's preparedness and history of communicable disease management, the country was tasked with addressing an unprecedented health care challenge.

As referenced earlier, and worth mentioning again here, the country of South Africa was in the process of transforming its health care system with the introduction of the National Health Insurance Bill when the COVID-19 pandemic began. This would have amended the National Health Act with the aim of providing universal health coverage via the introduction of a National Health Insurance scheme. While universal health coverage is a laudable goal, there were concerns that the reforms would overhaul the entire health care system in South Africa without addressing the most notable challenges facing the country—basic health care infrastructural needs and the maintenance of the health care system. This is especially true as recent assessments have shown that the quality of social infrastructure in South Africa is poorly maintained and not meeting demand—with the ratings of public hospitals and clinics declining [8]. At, the onset of COVID-19—poor social infrastructure needed to be restored at all costs, even more so when considering that the country was already ranked poorly in the 2019 Global Health Security Index (i.e., 49 of 89 countries).

The announcement of the first case of COVID-19 in South Africa, on March 4, 2020, came in this context with much attention via national broadcast television. The national broadcast meant that South Africans were generally aware of the situation, but the crisis rapidly grew more serious as cases rose across the world and the country confirmed its second case soon thereafter.

The country's strategy was to contain the spread of COVID-19 immediately. However, battling the emerging pandemic proved challenging given the already limited investment in health infrastructure; this situation was compounded by the institution of a nationwide lockdown at the onset of COVID-19 that sought to mitigate the transmission of the virus. While the need for a blanket approach in the form of a nationwide lockdown seemed defensible, the implementation of this control measure limited access to essential primary health care services offered by the City of Cape Town—holding harsh implications for the city's social-political economy. This also meant that provincial governments had to rapidly implement high-level changes in their approaches for providing health care. The provincial government oversaw the execution of the overall health strategy, as directed by the National Government, while the city of Cape Town was tasked with providing auxiliary services to help manage COVID-19 in local communities. This arrangement ultimately led to substantial levels of uncertainty and confusion throughout South Africa at the start of the COVID-19 pandemic, especially when considering that the demographics of each of the eight South African provinces differ dramatically.

The Western Cape Provincial Government laid out its approach for responding to the COVID-19 pandemic in a strategy document issued on April 30, 2020 [9]. This strategy reflected the complex nature of the public health emergency and the demanding response that it required. As detailed in this document, key elements of the response strategy relied on building capacities to manage the anticipated surge in COVID-19 cases. For this reason, the blanket approach of nationwide lockdown—an approach that ultimately sought to reduce the demand for health services by reducing the number of cases—seemed cogent. Other crucial parts of the response strategy included creating isolation rooms and areas for providing critical care (while maintaining essential health services and managing the current need for care in hospitals), strategies for transporting and discharging cases, and enhancing morgue capacities. The document also detailed the strategies to be used for following up on confirmed cases of COVID-19. The clinical management of COVID-19 cases included considerations such as diagnostic testing and triaging, admission, and medical care required for each individual confirmed to have COVID-19.

As time went on, caseloads grew, and it soon became clear that a lockdown would not be sustainable for a prolonged period of time. However, much was also learned about the virus and disease, which allowed for the development of additional protocols for managing COVID-19 in public settings. There was also a separate operation plan for business continuity, with a specific focus on comorbidities once risk trends were identified. All of these plans required that specific action triggers be analyzed on a continuous basis, which necessitated that other, nonessential services be deprioritized. For example, some public health services such as surveillance and risk communication efforts were highly prioritized, as the entirety of the response depended upon understanding the level of risk and the readiness of the population to mount a commensurate response that could successfully manage the spread of disease.

Private sector actors aided in the response by using their outstanding technological capacities to support the integration of health services as a means of increasing availability. A notable example of a successful public-private partnership was witnessed when a field hospital was constructed at the Cape Town International Convention Centre to support the management of COVID-19 cases. This facility is co-owned by private stakeholders and the city of Cape Town, which holds shares, and the Western Cape Government supplied the necessary equipment—securing more than 100 beds to offer COVID-19 treatments.

One year later, the number of reported COVID-19 cases in South Africa stood at 3.6 million, with a recovery rate of approximately 97%. While this high rate of recovery is a notable achievement that may be attributed to the development and implementation of pandemic response plans, all of these strategies largely neglected what came after the initial response to the virus—a concern acutely felt by many South Africans who were left living in a level of uncertainty. In this regard, one major shortcoming of the plans was their inability to inspire a sense of hope for a better future—that the city of Cape Town, the Western Cape Province, and the Republic of South Africa would recover and find themselves in a better position compared to when the pandemic began. In essence, many South Africans longed for more than a return to normalcy and a cure for a disease.

Cape Town's COVID-19 response

The COVID-19 response in Cape Town is one that is characterized by two diverging narratives—one that relates to the desperate need to save lives by rapidly scaling up public health interventions, and another equally important narrative related to ensuring that the livelihoods of individuals would not be destroyed as a result of blanket approaches, like the nationwide lockdown.

To address this dissonance, the provincial government forged ahead with a Hot Spot Strategy—a phrase that soon became synonymous with the Western Cape's response and management of COVID-19. This approach aimed at managing cases in areas experiencing high levels of disease spread, while allowing economic activity for those areas that were not experiencing high burdens of COVID-19. The hope for this strategy was that it would help channel resources where they were needed most, while also allowing for flexibility in the pandemic response. Initially, this strategy enjoyed little support and was not adopted immediately. As the blanket bans continued, however, and more people were pushed into poverty, the need to return to some state of normalcy became increasingly clear—especially as finding a "cure" seemed a distant goal. Because of this, blanket bans were gradually lifted and the Hot Spot Strategy was adopted by the national government. Ultimately, while a success, the Hot Spot Strategy was only effective because of its engagement of a wide array of stakeholders—all levels of government, the health care sectors (i.e., public and private), private sector actors, and civil society organizations.

The Hot Spot Strategy also resulted in other innovations that came to characterize the local response in Cape Town. Among these was the development of field hospitals. As discussed previously, the first field hospital, the Hospital of Hope, was developed at the Cape Town International Convention Centre through a public-private partnership and allowed for a large number of COVID-19 cases to be contained and cared for in one space. After this initial effort, an additional facility for the provision of intermediate care was developed within the metropolitan area, with support from the provincial government. As alluded to earlier, the local government also played a crucial role in these efforts and used its resources, such as community halls, to provide spaces for additional beds and later vaccination sites. Soon thereafter, the government was able to develop field hospitals and facilities in each region of the Western Cape Province. And, because of the flexible nature of the Hot Spot Strategy, these facilities were able to make better use of the limited resources available to them.

Another innovation that played an important role in the local pandemic response was repurposing community health workers to become COVID-19 field workers. These individuals were trained to proactively identify possible COVID-19 cases and refer them to local testing sites.[4] Should an individual test positive for COVID-19, they were then referred to small, community-based isolation facilities that allowed authorities to contain the spread of disease. Community health workers also helped to alleviate the burden on the health system by continuing to deliver routine services and medications to those who would have otherwise visited local clinics for prescriptions.

Other elements of the pandemic response in Cape Town included the provincial government developing a telemedicine initiative and telecommunication strategy. These efforts sought to proactively help individuals access health care—especially those who were understood to be more susceptible to infection and at a higher risk of severe COVID-19 disease.

As for the other narrative, due to the stark inequalities and socioeconomic divides that exist in the city, the local government, as the authority closest to the people, was required to consider many other matters beyond primary health. The national government intervened to provide a social relief grant that helped support these efforts; still, the funding provided did not match the effects already lived or seen in Cape Town. The city ultimately repurposed its community development budget to help close this funding gap and provide additional humanitarian relief and response efforts.

---

4. To support this intervention, the city of Cape Town, in collaboration with the provincial government, established 40 drive-through testing sites that could help support the rapid detection, containment, and treatment of COVID-19.

One component supported by these response efforts included embarking on a de-densification strategy to curb the spread of COVID-19. From March to June 2020, roughly 60% of the COVID-19 reported cases were detected in Cape Town's informal settlements. The densification process, therefore, aimed to address concerns regarding the potential for rapid disease transmission as a result of dense populations and the close proximity of living conditions that characterize these environments. R500 million were spent as a crucial first point of call. Importantly, residents were not removed from their homes because of this strategy, which instead, had broader aims of addressing the economic and social conditions that characterize these environments. And, while not the primary motivation of these efforts, they were also an important first step in recognizing the conditions that could help support broader well-being goals.

There were also offers presented on rate rebates and temporary payment arrangements to support those who were unemployed. As a part of these, the indigent threshold was raised to R7000 per month to allow a greater number of city residents to qualify for free services that were offered by the government [10]. The city also worked to enroll more people in these services by enabling a faster registration process for indigent, disabled, and pensioner rebates.

Other opportunities focused on addressing the societal impacts of the pandemic were afforded as a result of the city finding itself on the brink of a local election cycle during the COVID-19 pandemic. Because of this, the local government was presented with a unique opportunity to propose new pledges to its COVID-19 response and recovery plan. These pledges included a variety of considerations including improving the city's infrastructure, developing back-to-work programs, and enhancing facilities as a means of attracting local, domestic, and international business.

Still, these response efforts did not come without their challenges. Beyond the regretful loss of life associated with the pandemic, the hard-handed enforcement of regulations and the limited provision of public services took their toll on the social fabric. Early on, it appeared to many that the price to pay for a rapid response was government accountability—as there were numerous stories of corruption related to the acquisition of personal protective equipment from the National South African Health Department. This, while other unceasing concerns of fires, floods, and crime, continued unabated and threatened the livelihoods of individuals and families across the country.

If one had visited the country during the height of the pandemic, they likely would have been left with a bleaker view. In July 2020, South Africa faced a different state of emergency—mass looting in two provinces that resulted in roughly R3.5 billion lost [11]. One reason often cited as causing—or at least contributing to—these mass lootings is the devastating economic impact of the pandemic and the suffering caused by pandemic response interventions. Indeed, many found themselves without basic services and provisions as a result of supply chain issues that stemmed from the pandemic and it must be noted that, similar to many other parts of the world, the COVID-19 pandemic has exasperated the inequality gaps in South Africa [12].

While the actual reasons for the looting are likely more nuanced, one thing is for certain—there is a need for a holistic "cure" —not only one that prevents death and suffering from the disease but also one that alleviates social and economic losses as a result of the pandemic. As previously discussed, South Africa, broadly, and Cape Town, in particular, are known for their tourist attractions and tourism is a main driver of the local economy and job creation. While the development of vaccines and the possibility of a treatment were certainly warmly welcomed, there was also a need to address all of the other societal harms that resulted from the impact of COVID-19.

This is especially true when considering that even the vaccines were not immune to the plagues of logistics, distrust, and mis- and disinformation. With regard to vaccines, challenges in Cape Town have included everything from vaccine expiry dates to conspiracy theories on vaccine side effects, to addressing and overcoming a strong "anti-vaxxer" campaign. All of these issues have contributed to a sense of vaccine hesitancy—a challenge that demands a strategy of its own. In Cape Town, this hesitancy has worked in tandem with a general sense of apathy that has arisen due to how vaccines were rolled out and introduced to the population. While there was little planning that could have been done for the immediate response to the COVID-19 pandemic—especially given the novel nature of the disease—there was more than sufficient time to plan for and develop a system that would ensure the efficient, equitable, and rapid rollout of vaccine in the face of high global demand. While some degree of waiting was understandable, the substantial delays that have characterized the global vaccine rollout have reduced the urgency and vigor many feel toward the vaccine.

Plainly stated, what was needed alongside all of the detailed strategies that have dominated the pandemic response—those for tracking cases, reducing disease transmission, providing health care services, and managing deaths—was a plan for healing the societal traumas that are associated with COVID-19.

Conclusion

The response to the COVID-19 pandemic in Cape Town is one that, at surface level, may appear successful given the context in which it occurred and state of the health care system. The powers for health preparation and implementing pandemic response lay squarely within the ambit of the provincial governments, in this case, the Western Cape Government, following direction from the national government. And, throughout the pandemic, the local government in Cape Town was primarily responsible for supplying the routine medical services that were needed by the population—not those related to acute or emergency care, such as COVID-19. As such, as the world was brought to a standstill in 2020, the pandemic response in South Africa is one may be characterized involving strategic efforts from the national government, innovation and leadership from provincial governments, and implementation by local governments that moved to avail local spaces and resources to implement strategies, such as the Hot Spot Strategy. These efforts have, fortunately, resulted in an excellent recovery rate, and when considering the state of the health care system that had been left largely neglected by the national government before the pandemic, the response went reasonably well.

However, a lesser-spoken narrative that accompanies this success is that many in Cape Town and South Africa have a diminished sense of hope and pride—both of which are necessary for moving forward. Even those that have survived COVID-19 continue to endure great loss. Many individuals were leading financially and socially sustainable lives before the pandemic, only to lose everything—their jobs as businesses closed because of lockdowns, their homes as rents became increasingly unaffordable, their family members as routine medical services were not available, and their direction in life. Consequently, many feel a profound sense of fatigue toward COVID-19 and the extreme response measures it imposed on society.

Ultimately, the past several years have been defined by many challenges, but presently, Cape Town finds itself in a period defined by many questions. Perhaps none more important than "where to next?" Is this future one defined by resilience and hope? Is it one defined by greater equity? Or, have we already missed these deadlines?

To answer these questions, and to truly end this COVID-19 pandemic, it seems necessary to develop a continuum that portrays reflections on the past, as well as aspirations for the future. If done thoughtfully, this continuum will show a response that was defined by successes but also a shortcoming of failing to target and pursue the ultimate goal—a national recovery and instilling hope for all of those that have suffered trauma and loss. Only when this is achieved, the confusion that has defined the pandemic response can be eliminated, the opportunities once known can be re-enjoyed, and the courage to pursue a better life and future can be instilled in the hearts of South Africans.

References

[1]   Western Cape Government. Socio economic profile–Cape Town. Cape Town: Western Cape Government; 2020.

[2]   Githahu M. If Cape Town population continues to explode, the Western Cape may need a new city. Independent Online 2022;26 Oct.

[3]   Legg K. W Cape shows growth in its population. Independent Online 2013;6 Jun.

[4]   Government of the Republic of South Africa. The National Health Act 61 of 2003. Pretoria: Government of the Republic of South Africa; 2003.

[5]   Department of Statistics South Africa. Public health care: how much per person?, <https://www.statssa.gov.za/?p = 10548>; 2017 [accessed 23.11.22].

[6]   Lencoasa T, Mapipa K, Chaskalson J. Health needed a recovery budget — not more cutbacks. News24 2022;22 Mar.

[7]   Gilbert L, Walker L. Treading the path of least resistance: HIV/AIDS and social inequalities a South African case study. Social Science & Medicine 2002;54(7):1093−110. Available from: https://doi.org/10.1016/s0277-9536(01)00083-1.

[8]   Burger S. South Africa's infrastructure quality rated D, social infrastructure deteriorating. Engineering News 2022;11 Nov.

[9]   Western Cape Government. COVID-19 risk adjusted strategy. Cape Town: Western Cape Government; 2020.

[10] Democratic Alliance. "2020 Cape Town Council expands indigent relief, passes SA's lowest rates increase." Democratic Alliance; 27 May 2020.

[11] VOA News. "Poverty at root of South Africa violence and looting: analyst." VOA News; 19 Jul 2021.

[12] Yozan N, Lakner C, Mahler D. Is Covid-19 increasing global inequality? World Bank Data Blog 2021;7 Oct.

Chapter 11

A proactive approach to curtail the spread of COVID-19 in the Dharavi Slum, Mumbai, India

Parangimalai Diwakar Madan Kumar[1], Saravanan Poorni[2] and Zameer Shervani[3]

[1]Department of Public Health Dentistry, Ragas Dental College and Hospital, Chennai, Tamil Nadu, India, [2]Department of Conservative Dentistry and Endodontics, Sri Venkateswara Dental College and Hospital, Chennai, Tamil Nadu, India, [3]Food and Energy Security Research and Product Centre, Sendai, Japan

Background

Dharavi is a slum located in Mumbai, India that is recognized as one of the densest slums in Asia and one of the largest in the world. Along with an ever-increasing population density, low incomes, small and cramped homes, and a dearth of hygiene and sanitation services have contributed to poor living conditions in the Dharavi Slum. When cases of COVID-19 were reported in the slum in early April 2020, controlling the spread of the disease was an immense challenge as observing and implementing response measures such as physical distancing in an overpopulated slum area was very difficult. In this chapter, we will discuss the efforts taken to control the spread of disease in the Dharavi Slum.

Background on Mumbai, India

Mumbai is situated on the Western Konkan coast of India and is the most populous metropolitan city in the country.[1] It harbors a population of close to 12.5 million individuals based on the 2011 Census of India [1]. However, the Mumbai Metropolitan Region is much larger and is estimated to hold a

---

1. The city was formerly known as "Bombay"—an anglicized name used by the British when they took control of the city in the 17th century—but was renamed to "Mumbai" in 1995 in an effort to shed the legacy of European colonialism.

Inoculating Cities, Volume II: Case Studies of the Urban Response to the COVID-19 Pandemic.
DOI: https://doi.org/10.1016/B978-0-443-18701-8.00013-8

population of at least 20 million [1]. For the sake of decentralized adminis-tration, the city is divided into 24 wards under Municipal Corporation of Greater Mumbai (MCGM). The literacy rate in Mumbai is around 89.73% (92.56% for males and 86.39% for females) [1]. The city is the "Financial Capital" of India, as it contributes to 6.16% of India's gross domestic prod-uct, 25% of industrial output, 70% of maritime trade in India, and 70% of capital transactions to India's economy [2]. It is also the home of 38 billio-naires—the eighth-highest number of billionaires in any city in the world—and the highest number of millionaires and billionaires in India [3]. And in 2008, Mumbai's billionaires had the highest average wealth around the world.

Despite this wealth, the city is also characterized by extreme socioeco-nomic inequality. Mumbai is home to a large homeless population of people who are unable to access housing or proper places to live in. Further, accord-ing to a report released by the Centre for Science and Environment, 3.6% of the city's population lives below the poverty line [4], and an estimated 55% of Mumbai's population lived in slums [5]—defined as a densely populated area typically characterized by poverty, a deteriorating built environment, and poor living conditions. Mumbai is estimated to have the largest slum population of any city in Asia.

In India, each state uses different criteria to determine if a community qualifies as a legal slum, which is a prerequisite to the receipt of municipal services such as piped water, toilets, electricity, or public transportation [6]. Because of this, a *de facto* slum may not secure *de jure* status resulting in two types of slums: notified slums (i.e., those recognized by the government) and non-notified slums (i.e., those that are not recognized by the govern-ment) [6]. Almost half of Mumbai's slums are non-notified. Still, of those that are notified, the largest slum in Mumbai is called Dharavi. An estimated 850,000 people live in nearly 55,000 dwelling units in Dharavi, with a popu-lation density of 340,000 people per square kilometer [5]. However, and importantly, while densely populated, not everyone in the slum lives below the poverty line. The Dharavi Slum is also home to a middle-class, 15,000 single-room factories, and some 5000 businesses.

The public health system in Mumbai

Per the Mumbai Municipal Corporation Act of 1888, providing primary health care is an obligatory duty of the corporations of Mumbai [7]. The main focus of primary health care is to strengthen preventive measures by supporting health infrastructure, ensuring an adequate health care workforce in dispensaries and hospitals, and maintaining a strong disease surveillance mechanism.

Health services for people in Mumbai are provided through a network of hospitals and dispensaries run by the MCGM, the Maharashtra State

Government, and private sector actors. The MCGM maintains 4 medical college hospitals, 1 dental college hospital, 16 municipal general hospitals, 6 specialty hospitals (i.e., tertiary hospitals), 29 maternity homes, 175 municipal dispensaries, and 183 health posts. Additionally, the Maharashtra State Government has one medical college hospital, three general hospitals, and two health units located in Mumbai [8]. The medical college hospitals and the specialized hospitals provide basic medical services, as well as tertiary-level care, while peripheral hospitals are designed to provide primary and secondary care. At these facilities, serious illnesses are recognized and further referred to one of the six specialty hospitals where individuals may receive more specialized health care. The MCGM hospitals are also involved in an array of national health programs such as the National Leprosy Eradication Program, the National Malaria Control Program, the National Program for Family Welfare and Maternal Child Health, the National AIDS Control Program, and the National School Health Program. These aforementioned hospitals maintain an indoor bed capacity of greater than 12,000 beds.

This network of health care facilities also includes close links to more community-oriented facilities, such as dispensaries and health posts. These facilities have been established to provide convenient and easy-to-access health care for people seeking treatment for diseases and ailments that do not require a visit to a major hospital. Still, they are linked to the peripheral hospitals in their respective wards should an individual require more specialized medical care. Typically, 15 million patients are treated annually in outpatient departments located in the hospitals and dispensaries managed by the MCGM.

For economically disadvantaged people, however, access to health care is not a given. To remedy this challenge, recent years have seen a variety of government-sponsored efforts to provide greater access to health care services, such as the Rajiv Gandhi Jeevandayee Arogya Yojana—a universal health care and insurance scheme run by the Maharashtra State Government, and Janani Shishu Suraksha Karyakram—a national initiative that provides free and cashless services to pregnant women.

COVID-19 in India

While central government agencies in India had experience responding to previous health emergencies, such as the Nipah virus and SARS, the magnitude of the COVID-19 outbreak was unprecedented. The first case of COVID-19 in India was reported on January 30, 2020. The index patient was a student who had returned from Wuhan, China and only two more cases were reported thereafter during February.

However, additional cases were subsequently identified and reported in March, followed by a large surge in the number of cases in the latter half of April 2020. By June 9, 2020, the Ministry of Health and Family Welfare reported a total of 266,598 confirmed COVID-19 cases which were reported from all the states and union territories in India [9].

Among all Indian states, Maharashtra was the most impacted. The state detected its first case on March 9, 2020 in an individual that lived in Pune, who had recently returned to the country from Dubai. And, of importance to this chapter, the first case of COVID-19 in the Dharavi Slum was reported on April 1, 2020. An epidemiological investigation suggested that this index case contracted the infection from the Tablighi Jamaat congregation, a religious gathering that occurred at the Nizamuddin Markaz Mosque in Delhi in early March 2020. This event—attended by nearly 9000 missionaries from various parts of India, as well as 960 participants from some 40 South East Asian countries—was later identified as a super-spreader event, accounting for over 4200 confirmed cases and at least 27 deaths associated with COVID-19 infection.

The response to COVID-19 in the Dharavi Slum of Mumbai

Despite being the richest city in India, and receiving assistance from the central government, MCGM's public health authorities faced a daunting challenge in responding to the COVID-19 outbreak [8]. By the end of April 2020, one month after the first case of COVID-19 was reported in the slum, 491 cases had been confirmed in the slum, with a 12% growth rate and a case-doubling period of 18 days [5].

As such, a strategy geared toward containing transmission was developed and implemented. The strategy, named "Chase the Virus," drew a foundation from the "Four Ts": tracing, tracking, testing, and treatment. The strategy ultimately sought to rapidly reduce the number of cases, while providing treatment to those who were infected with the virus. More specific details of this strategy and the control methods used are described hereafter.

Administrative efforts

The MCGM imposed a strict lockdown and closed the border of the slum, with 24 posts at all entry-exit points to monitor the movements of slum residents. Additionally, strict containment measures were used in virus hotspots, including the use of drones to monitor the movement of persons confirmed to be infected. In the event that individuals did not comply with quarantine and isolation regulations, the drones alerted the police who then intervened to enforce the infection control measures. Groups of volunteers were also employed to assist and monitor the process of isolation and quarantine.

While impactful from a public health perspective, the stringent lockdown also resulted in significant economic hardship for the Dharavi Slum residents and was debilitating for daily wage earners and migrant laborers. Most of these individuals had little or no financial savings and were left with difficult decisions regarding complying with lockdown measures or breaking them to procure daily necessities, such as food and other essential items. To address this challenge and limit the need to break lockdown protocols, the State government and MCGM worked together to ensure the regular, door-to-door delivery of items and daily necessities, such as milk, vegetables, groceries, and other essential goods. Importantly, these efforts strove to be culturally considerate. For instance, the authorities worked with local nongovernmental organizations (NGOs) and made special efforts to provide the Muslim Brotherhood of the Dharavi Slum with a supply of fruits and seviyan that would enable them to break their fast during the holy month of Ramadan.

Further, following the announcement of a nationwide lockdown on March 24, 2020 [10], approximately 100,000 migrant workers and their families left the slum. Their departure arrangements were supported by the Central and State administration, while giving due consideration to other health-related issues. Still, while considerations were given to the potential health implications of this population movement, this exodus may have resulted in the spread of the virus to other parts of the country.

Screening and surveillance efforts

The Maharashtra State Government deployed nearly 2500 doctors and health care workers from Mumbai's public health system to the Dharavi Slum to support the local pandemic response efforts. This public workforce was supplemented by the MCGM, which recruited an additional 350 private practitioners. This health care workforce was critical for screening and surveillance efforts and helped the MCGM public health staff in identifying the symptoms of infected persons, referring individuals for medical treatment, and advising both confirmed cases and healthy slum residents on how to mitigate the health impacts of the outbreak.

The MCGM also opened several fever clinics in the slum to aid in the screening and testing of slum residents that presented with symptoms consistent with influenza-like illness. At these clinics, residents were checked and re-checked with thermal scanners and oximeters, and any individual with symptoms was sent for additional diagnostic laboratory testing. Individuals confirmed to be infected were then isolated at facilities prepared by the local civil authorities. Within a few weeks, by the first week of May 2020, municipal teams had screened 548,270 people in the Dharavi Slum, including some 120,000 senior citizens who were recognized to be more vulnerable to severe disease.

Quarantine and isolation

While surveillance efforts represented one important response measure, equally important was providing services to suspected and confirmed

COVID-19 cases. To this end, establishing large quarantine centers was also an important measure for the pandemic response. The MCGM and state government established a total of nine quarantine facilities in the Dharavi Slum by converting schools, marriage halls, and community centers into temporary health facilities. These facilities were equipped to provide basic essentials, regular checkups, and free health care and testing. The Rajiv Gandhi Sports Complex was the first large facility that was requisitioned by authorities and contained 300 beds to help support COVID-19 isolation and quarantine efforts.

In total, some 9500 Dharavi residents were placed under quarantine for 2 weeks [11]. Due to social isolation and the uncertainties surrounding the infection, people who were quarantined and isolated faced deteriorating mental health. To address this crisis, the administration organized yoga, aerobics, and breathing exercise sessions to ameliorate the heightened stress and anxiety levels felt by those in quarantine.

To give medical support to COVID-19-infected patients, especially high-risk individuals, the local civil administration also took control of five private hospitals along with their staff and medical equipment. This was made possible by evoking the National Disaster Management Act of 2005 by the Government of India. Through this action, it became mandatory for all private health providers to handle and treat COVID-19 cases based on guidelines set forth by the government, as well as to report the clinical outcome of the treatment to the national surveillance authorities.

Public-private partnerships

As previously referenced, through the National Disaster Management Act of 2005, the MCGM requested that all local privately practicing doctors support the COVID-19 pandemic response in the Dharavi Slum by opening their clinics. To incentivize health care workers to support response efforts in the slum, authorities provided personal protective equipment (PPE) kits (i.e., gloves, gowns, and N95 respirators) to all health care workers. These kits contained a 7-day supply of PPE that was necessary for the safety of the health care workers while attending the infected patients; the supplies were restocked after 1 week.

In total, 350 local practitioners agreed and worked as a bridge between municipality officials and the people living in the slum. As previously discussed in greater detail, with the help of these individuals, authorities established temporary fever clinics and medical camps to screen high-risk and vulnerable individuals. Irrespective of symptoms, all patients presenting to these clinics were screened for fever and their oxygen saturation levels were checked.

Ultimately, this public-private partnership approach yielded rich dividends and helped in rapidly implementing the "Chase the Virus" strategy to control the spread of COVID-19 in the Dharavi Slum. This is, in part, because the residents of the slum trusted the doctors, many of whom lived in or close to the community—leading to faith in their clinical prognosis and diligent adherence to instruction.

Community participation

Recognizing the importance of involving the community in the pandemic response, the MCGM worked to identify and collaborate with influential local leaders, community organizations, and NGOs. The administration recruited hundreds of volunteers from the community and designated them as "COVID Warriors," before asking them to help ensure that slum residents had regular and sustained access to daily essentials such as groceries and medicine in the priority containment zones.

Other noteworthy efforts include involving community religious leaders in risk communication activities. For instance, during the initial days of the national lockdown, during every namaz, religious leaders explained to people how important it was to take proper precautions against COVID-19. Many maulvis and maulanas also carried photographs of the deceased during their visits to the slum to help residents understand that the threats posed by the virus were real and bear witness to the havoc and destruction that COVID-19 caused for so many families.

Conclusion

The public health measures and proactive strategy adopted by the MCGM helped to reduce the burden of COVID-19 in the Dharavi Slum. While the COVID-19 growth rate stood at 12% at the end of April 2020, this was reduced to 4% in May 2020, and to 1% by June 2020. Similarly, the case-doubling period stood at 18 days in April but had improved to 43 days in May, and 78 days in June. In early July, there were a total of 2335 confirmed cases in the slum, with a tally of 1735 recoveries (recovery rate 74%). Further, this recovery rate may be much higher in reality as, despite all the response efforts taken, a serological survey conducted between July and August 2020 suggested that 57% of the total population living in the slum had been infected by the virus [12].

Still, it appears that the "Chase the Virus" Strategy and actively following the "Four Ts," enabled authorities to successfully flatten the COVID-19 curve in the Dharavi Slum within 2 months. And various steps taken by the government and policymakers to implement this response strategy in the slum were subsequently used in other locations in India. This is noteworthy, as while many other countries with advanced health systems faced difficult challenges in containing the COVID-19 pandemic, these challenges were acutely felt in a context defined by extremely limited resources.

Accordingly, the pandemic response approach used to respond to COVID-19 in the Dharavi Slum—one that relied on prompt and visionary leadership, proactive thinking, efficient governance, public-private partnerships, and community engagement—may offer lessons for other poor, densely packed urban slums located around the world, and represent a compelling model for responding to outbreaks in resource-limited settings.

References

[1] Government of India − National Informatics Centre. India statistics: million plus cities in India as per Census 2011, <census2011.ac.in>; 2011 [accessed 02.12.22].

[2] Reserve Bank of India. Department of Statistics and Information Management. Handbook of Statistics on Indian Economy. Kolkata: Reserve Bank of India; 2022. p. 2021−2.

[3] Tognini G. World's richest cities: the top 10 cities billionaires call home. Jersey City: Forbes Magazine 2020;.

[4] Narainn S, Mahapatra R, Das S, Misra A, Banerjee S. State of India's environment 2022: inf. New Delhi: Centre for Science and Environment; 2022.

[5] Kaushal J, Mahajan P. Asia's largest urban slum-Dharavi: a global model for management of COVID-19. Cities 2021;111:103097. Available from: https://doi.org/10.1016/j.cities.2020.103097.

[6] Subbaraman R, O'Brien J, Shitole T, Shitole S, Sawant K, Bloom DE, et al. Off the map: the health and social implications of being a non-notified slum in India. Environment and Urbanization 2012;24(2):643−63. Available from: https://doi.org/10.1177/0956247812456356.

[7] Mahapatra P, Khetrapal S, Nagarajan S. An assessment of the Maharashtra State Health System. Mandaluyong: Asian Developmental Bank 2022;. Available from: https://doi.org/10.22617/WPS220063-2.

[8] Municipal Corporation of Greater Mumbai. Health services background: Mumbai. <https://portal.mcgm.gov.in>; 2022 [accessed 02.12.22].

[9] Pal R, Yadav U. COVID-19 pandemic in India: present scenario and a steep climb ahead Journal of Primary Care & Community Health 2020;11:2150132720939402. Available from: https://doi.org/10.1177/2150132720939402.

[10] Press Information Bureau, Government of India. 24 Mar Government of India issues Orders prescribing lockdown for containment of COVID19 epidemic in the country. New Delhi: Government of India; 2020.

[11] Ministry of Health and Family Welfare, Government of India. Guidelines for quarantine facilities COVID-19. New Delhi: Government of India; 2020.

[12] Shervani Z, Bhardwaj D, Nikhat R. Dharavi Slums (Mumbai, India): the petri dish of COVID-19 herd immunity. European Journal of Medical and Health Sciences 2021;3(3):38−41. Available from: https://doi.org/10.24018/ejmed.2021.3.3.860.

Section V

Risk communication

What is risk communication?

The World Health Organization (WHO) defines risk communication as, "the real time exchange of information, advice, and opinions between experts, community leaders, or officials and the people who are at risk" [1]. There are two key subdivisions to emergency risk communication: internal communication between officials, and external communications with the general public [2]. During a public health emergency, internal communication between experts and officials is needed to coordinate the response and consistent public messaging across departments/officials. External communications are crucial to deliver information about the health risks they face and provide evidence-based recommendations for how they can protect themselves [1].

The International Health Regulations (IHR 2005) require states to establish sufficient capacity to detect and report a potential public health emergency of international concern to the WHO within 48 hours [3]. Public health officials must therefore have established internal communication systems to connect local health authorities to national health authorities. More generally, risk communication is the sixth of eight core capacities WHO member states are required to develop under the IHR at the community, intermediate, and national levels.

What is good risk communication?

Good risk communication starts early, is culturally sensitive, based on scientific evidence, and transparently communicates levels of uncertainty while providing clear recommendations [1,4]. Communicating uncertainties and adjusting guidance to align with the lived experiences of communities were necessary to maintain trust and compliance during COVID-19 [1,4]. For example, contradictory and confusing recommendations on the wearing of facemasks from the WHO and other institutions early in the pandemic eroded public trust [5–7]. Finally, the digital ecosystem has facilitated the spread of misinformation, particularly on social media. When done well, government authorities monitored these trends and addressed what they were seeing—often using social media platforms themselves to counteract the spread of rumors/misinformation and expand their audience [1].

Why does risk communication matter?

External risk communication is fundamental to an outbreak response because it gives the public the ability to make informed decisions about their health and safety. For example, people get vaccinated and wear masks to prevent transmission of a respiratory virus or use protection to prevent transmission of a sexually transmitted disease, but only if they are made aware of the risks and effective prevention strategies. It also helps facilitate compliance with non-pharmaceutical interventions that require large-scale cooperation to be effective, like masking and lockdowns.

Local public health officials rapidly adapted risk communication to known endogenous sociocultural factors that affected emergency response, whereas national risk communication was broader and more slow-moving. Further, although the majority of pandemic response activities were overseen by national authorities, the responsibility for implementing effective risk communication fell primarily with local governments [8]. Local authorities were particularly suited to this challenge because of their established relationships with the communities they served. This relationship allowed for heightened levels of trust and provided crucial background knowledge on how to best serve and protect various subpopulations [8]. For instance, urban health officials must modify outreach strategies and materials to reach all facets of the diverse populations that call their city home. This is often a daunting task. The population of New York City, for example, speaks more than 200 languages, and identifying appropriate channels and languages to reach all segments of such a diverse population proved difficult during the COVID-19 response [9].

Chapter introduction

The following chapter outlines how Baltimore City, Maryland, United States of America engaged in risk communication activities (Fig. 1). Internally, Baltimore used its preexisting incident command system to successfully manage communications between various offices and departments. While, externally, the city tailored external communications to the local population designed its information campaigns to encourage citizens to wear facemasks, practice social distancing behaviors, and get vaccinated once vaccines became available. Baltimore City is a case study in clear and concise messaging, targeted community outreach, and even humor to conduct external outreach uniquely tailored to meet the needs of a diverse urban population.

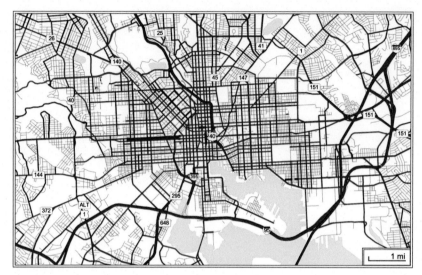

FIGURE 1 A map of Baltimore, Maryland, United States of America. Base map sourced from OpenStreetMap, using materials made available under CC BY-SA 2.0.

References

[1] World Health Organization. Communicating risk in public health emergencies: a WHO guideline for emergency risk communication (ERC) policy and practice. Geneva: World Health Organization; 2018.

[2] Zhang L, Li H, Chen K. Effective risk communication for public health emergency: reflection on the COVID-19 (2019-nCoV) outbreak in Wuhan, China. Health care 2020;8(1):64. Available from: https://doi.org/10.3390/healthcare8010064.

[3] World Health Organization. International Health Regulations (2005). 3rd ed. Geneva: World Health Organization; 2016.

[4] Cairns G, de Andrade M, MacDonald L. Reputation, relationships, risk communication, and the role of trust in the prevention and control of communicable disease: a review. Journal of Health Communication 2013;18(12):1550−65. Available from: https://doi.org/10.1080/10810730.2013.840696.

[5] Jingnan H. Why there are so many different guidelines for face masks for the public. NPR 2020;10 Apr.

[6] Lai-yam Chan A, Leung CC, Lam TH, Cheng KK. "To wear or not to wear: WHO's confusing guidance on masks in the covid-19 pandemic.". BMJ Opinion 2020;11 Mar.

[7] Kim DKD, Kreps GL. An analysis of government communication in the United States during the COVID-19 pandemic: recommendations for effective government health risk communication. World Medical & Health Policy 2020;12(4):398−412. Available from: https://doi.org/10.1002/wmh3.363.

[8] Boyce MR, Asprilla MC, van Loenen B, McClelland A, Rojhani A. Urban pandemic response: Survey results describing the experiences from twenty-five cities during the COVID-19 pandemic. PLoS Global Public Health 2022;2(11):e0000859. Available from: https://doi.org/10.1371/journal.pgph.0000859.

[9] Dorn S. "Language barriers compound COVID-19 challenges among non-English speakers.". City & State NY 2022;24 Jan.

Chapter 12

Risk communications in a pandemic: tailored communications and public health messaging to urban populations in Baltimore City

Letitia Dzirasa[1], Adam Abadir[2], Jennifer Martin[3] and Kimberly Eshleman[4]

[1]Baltimore City, Office of the Mayor, Baltimore, MD, United States, [2]Office of the Comptroller, State of Maryland, Annapolis, MD, United States, [3]Baltimore City Health Department, Baltimore, MD, United States, [4]Office of Public Health Preparedness and Response, Baltimore City Health Department, Baltimore, MD, United States

Background

According to 2020 data from the decennial census conducted by the United States Census Bureau, Baltimore City, Maryland's population size of 585,708 ranks fifth in the nation of predominantly Black cities. Approximately 64% of the city's population is Black, while 29% of the population is White and about 5% of the population is Latinx.

The population's current hypersegregated geographic distribution across the city has been largely influenced by the historical practice of redlining. In 1910 Baltimore City enacted one of the first pieces of "redlining" legislation in the country, promoting the racial segregation of neighborhoods [1]. In accordance with this legislation, housing developments, and neighborhoods set up restrictive covenants that refused the admittance of Jewish or Black residents [1]. These covenants effectively prevented communities of color from building wealth and established segregated neighborhoods, wherein individuals could be deprived of school funding, loans, public utilities and services, and other resources without explicitly citing race as a factor [2]. The result of these policies is racially segregated neighborhoods in Baltimore that are directly linked to key racial inequities in socioeconomic mobility as

Inoculating Cities, Volume II: Case Studies of the Urban Response to the COVID-19 Pandemic.
DOI: https://doi.org/10.1016/B978-0-443-18701-8.00015-1

well as physical, environmental, and social risks that negatively affect health [3]. The segregation of neighborhoods by race is also reflected in the rates of poverty across the city with all but three of the city's census tracts with poverty rates above 35% being located within predominantly Black neighborhoods. Conforming to national trends, these segregated neighborhoods suffer from worse health outcomes than majority-white neighborhoods. For instance, there is approximately a 20-year difference in life expectancy between neighborhoods in Baltimore City. According to the Johns Hopkins Urban Health Institute, in Baltimore City, "every single health indicator, including diabetes, obesity, high blood pressure, childhood asthma, smoking, and poor mental health days is higher for Black than for white residents... Of greatest concern is the substantial disparity between childhood asthma rates at 38.16% (Black) and 11.36% (white), more than a threefold difference" [2].

The Baltimore City Health Department (BCHD) is the longest continuously operating health department in the United States. Established by the Governor of Maryland in 1793 in response to a yellow fever outbreak, BCHD operates under the city's charter. Unlike other health departments in the state of Maryland, the BCHD and its staff are employed by the Baltimore City government. Executive leadership for the BCHD includes a Commissioner of Health who is appointed by the Mayor and subsequently confirmed by the Baltimore City Council. Pursuant to the Baltimore City Charter, the BCHD holds the authority to implement public health mitigation measures for the purposes of protecting the health of city residents. Per Article VII, §56 of the Baltimore City Charter, the Department of Health's powers and duties are as follows: BCHD shall (1) cause all laws for the preservation of the health of the inhabitants of Baltimore City to be faithfully executed and exercise those other powers and perform those other duties that are prescribed by law, (2) establish and implement policy for the treatment and prevention of physical and mental illnesses and for the education of the public with respect to environmental, physical and mental health, and (3) have general care of, and responsibility for, the study and prevention of disease, epidemics, and nuisances affecting public health.

While the BCHD operates independently of the Maryland Department of Health (MDH), the state's Department of Health, the BCHD has very close ties and alignment with the work being done at the state's health department. The BCHD works in partnership with MDH to protect the health and well-being of city residents. The MDH serves as a significant funder for many core public health services provided by the BCHD. Additionally, the MDH provides guidance around public health mitigation measures including support for schools, daycares, and congregate settings.

Additionally, the BCHD partners with local health care partners including the 11 acute care hospitals in Baltimore and the five federally qualified health care centers in Baltimore. Some of the work done with local health care entities and community-based organizations (CBOs) is coordinated through the BCHD's Local Health Improvement Coalition (LHIC). Baltimore's LHIC serves as a group of jurisdictional-level stakeholders that set broader public health priorities for the community. This group is led and coordinated by the BCHD and its current areas of priority include care coordination, diabetes prevention, and the social determinants of health. In addition to the health care partners that support the LHIC, CBOs play a key role in ensuring priorities and strategies are community informed with a specific focus on health equity.

For over two decades, the BCHD has had a public health preparedness program that engages in public health preparedness planning, training, exercises, and response. Plans are updated on a routine basis and corrective actions are made based on lessons learned following exercises and real-world events. The BCHD exercises its plans for receiving, distributing, and dispensing medical countermeasures and supplies routinely as part of its funding requirements under the United States Center for Disease Control and Prevention's (CDC) Cities Readiness Initiative and Public Health Emergency Preparedness grants.

Prior to the COVID-19 outbreak, the BCHD had a pandemic influenza plan that had last undergone major revisions following the 2009−2010 H1N1 influenza pandemic. The BCHD's Office of Public Health Preparedness and Response (OPHPR) leads the agency's coordination and planning for public health emergencies and works with each division to update its continuity of operations plans annually. In January 2020, as COVID-19 cases were being identified globally, the BCHD OPHPR reviewed the department's pandemic response plan and used the agency's emerging infectious disease plan to respond to the new and emerging threat posed by COVID-19. Individual divisions within the BCHD were tasked with updating their continuity of operations plans and helped to inform the pandemic response plan, which is regularly evaluated to ensure optimization of the response. Throughout the COVID-19 pandemic, the BCHD has performed mid-action and after-action reviews to identify strengths and opportunities for growth during the response.

In January 2020, when the World Health Organization (WHO) declared the COVID-19 outbreak a Public Health Emergency of International Concern and the United States Department of Health and Human Services declared a public health emergency, the BCHD partially activated its incident command system (ICS) in preparation for the city's response. The ICS is a flexible yet standardized structure to provide command, control, and coordination of resources toward common objectives. It seeks to establish clear lines of communication and can be used for both emergent incidents and planned events from the time an incident occurs until the requirement for management and operations no longer exists [4].

The BCHD routinely stands up as an ICS to manage public health emergencies and ensure the continuity of health and medical services during severe weather events, disease outbreaks, and other emergencies. In the 12 months prior to the response to COVID-19, the BCHD fully activated its ICS several times. Immediately before the pandemic, BCHD used ICS to support a tuberculosis investigation that involved a significant field screening and testing operation. In May 2019, the BCHD also activated ICS in response to a malware attack on the City's information and technology (IT) infrastructure to ensure essential services of the health department were maintained. Additionally, the ICS was activated for a small measles outbreak in April 2019. Laboratory-confirmed cases during this outbreak were limited to six total in the Baltimore metro region. However, during the BCHD's response to the measles outbreak, two mass vaccination point of dispensing clinics were run which allowed for the vaccination of 805 people. The second measles mass vaccination event occurred on the same day that the city's malware incident was discovered. Each of these responses leading up to the pandemic gave the BCHD staff practice and experience that was immensely valuable for the COVID-19 response.

The BCHD's ICS was fully activated in early March 2020. At the municipal level, a Unified Command structure was also established in March 2020. An initial tabletop exercise with city agency leaders reviewing the pandemic response and the necessary plans for coordination between agencies was done the first week in March, prior to the detection of the first cases in Baltimore or the state of Maryland. The first confirmed COVID-19 case was detected in Baltimore on March 14, 2020. During the early stages of the pandemic, identifying cases of COVID-19 was challenging due to a lack of adequate testing resources. An initial goal in the response was the establishment of a testing site to accommodate the anticipated demand for testing and the need to identify COVID-19-positive individuals for the purposes of contact tracing and subsequent isolation and quarantine, as necessary. The BCHD worked with Baltimore City municipal leadership, the MDH, the National Guard, and LifeBridge Sinai Hospital to establish the city's initial stationary testing site on the grounds of LifeBridge Sinai Hospital in Northwest Baltimore, a part of the 21215-zip code (Fig. 12.1). This test site, established in April 2020, was located in the city's first informally designated COVID-19 "hotspot." Though testing was open to anyone, it was notable that by the time this site was established, there was a massive uptick in cases in the 21215-zip code, which boasts a population that is 81.5% black [5].

Subsequently, in early April 2020, the BCHD began hearing anecdotally about an increase in cases among the Latinx population in Southeast Baltimore, in the 21224-zip code. As the BCHD looked to equitably increase testing access, the 21224-zip code, which houses a growing Latinx population, was one of a few sites selected for the first mobile testing locations. As testing locations increased, the BCHD posted information on hours and accessibility online via social media and the BCHD's website. Communicating the location of testing was critical to matching the resources

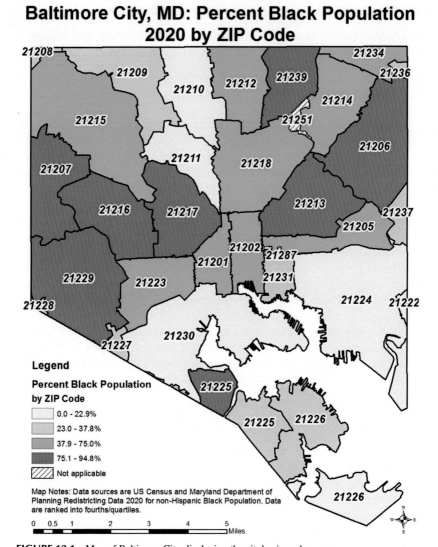

FIGURE 12.1 Map of Baltimore City displaying the city's zip code areas.

available with those communities most in need. What was equally important was clear messaging around the selection of testing locations. Those selections were based on both data indicating increased cases and information received directly from partners and residents of Baltimore. Approaching the COVID-19 response using a health equity lens drove decision-making regarding where to place limited resources including our messaging strategy and, in turn, built trust within the community.

COVID-19 risk communication efforts in Baltimore

Establishing initial communications: transparency, trust, and tailored messaging

As a local health department, resources to support communications and broad public health campaigns are generally very limited. Therefore, it was critical for the BCHD to be creative, innovative, and targeted in its messaging to reach the diverse residents of Baltimore City. The BCHD needed to have maximum impact with minimal expenditure of resources, especially at the beginning of the pandemic response, before funding from the federal government had been approved or awarded. Subsequently, the communication approach was initially centered around press conferences featuring local leadership including the Mayor of Baltimore and the BCHD's Commissioner of Health.

Baltimore's first press conference on COVID-19 was held on February 5, 2020—five days after the US Secretary of Health and Human Services, Alex Azar declared the outbreak a public health emergency. This initial press conference would be the BCHD's first effort at establishing trust with the community when it came to knowledge regarding the SARS-CoV-2 virus, the most recent data, and the overall response the city would use to reduce COVID-19 transmission. That first message was focused on what was known about the novel SARS-CoV-2 virus and what city leadership was doing to address the risks of a larger outbreak in Baltimore City. More specifically, the first message covered the following in detail:

1. Information about the virus that was known at the time, specifically that the virus was novel but familiar as a member of the coronavirus family of viruses;
2. Existing recommended preventative measures to reduce the risk of spread based on knowledge of other coronaviruses;
3. A historical review of where the virus started and who had been impacted to date;
4. A review of the ongoing involvement of the various city agencies and city leaders who were coordinating and collaborating on what a city-wide response would entail;
5. A review of the current threat level and what the emergency declaration by the CDC meant.

On March 5, 2020, the BCHD, in conjunction with the Mayor's Office, held another press conference preemptively as the BCHD anticipated cases would soon be detected in the state of Maryland. This would be the second of many more press conferences to come as the department's suspicions proved to be true—the next week, the outbreak was characterized as a global pandemic by the World Health Organization, and on, and the first case of

COVID-19 was identified in Baltimore City, on March 14, 2020. This second press conference allowed the BCHD to reiterate what was known about preventative measures, as well as how the city was preparing to respond, including activation of the City's Emergency Operations Center, the City's Joint Information Center, and the full activation of the BCHD's ICS. On March 15, 2020—in a press conference that assumed a graver tone, as the threat of COVID-19 had now become much more tangible—the BCHD announced the very first case of COVID-19 identified in a Baltimore City resident. During this press conference, the BCHD began to introduce case investigation and contact tracing (CI/CT) as critical components of the BCHD's role and responsibility. At the time, many were unfamiliar with what CI/CT was so early communications and explanations of CI/CT reassured the public that the BCHD had the expertise and in-house knowledge to evaluate cases. This would be the beginning of establishing the BCHD's role not just as a messenger but also as a key respondent to the needs of City residents during the pandemic.

The BCHD recognized, early on, that a key part of keeping residents safe would be empowering them to make informed decisions by ensuring they had the information necessary for making such decisions. Thus these initial press conferences proved to be vitally important, as the BCHD established its role as a trusted source of information and the preeminent local authority for providing the residents of Baltimore with the information necessary for preventing disease, keeping people healthy and safe, and understanding how the city was operating to address the imminent threat.

Tailoring the message

As the disease spread and epidemiological trends became more apparent, it soon became clear that the outbreak was having disparate impacts on older adults and minority populations. These developments justified more focused communication efforts. An early emphasis was then placed on attempting to communicate with specific affected populations, with a particular focus on understanding and acknowledging the limitations of public health communications and outreach efforts, based on community members' historic mistrust of public health. Early pandemic messaging focused on introducing and following harm reduction recommendations to protect the families and loved ones of Baltimore City residents from disease transmission (Fig. 12.2).

While messaging was shared over various social media accounts operated by the Health Department and the Mayor of Baltimore, print flyers were also created and distributed throughout the city, focusing on highly visible areas where targeted populations were most likely to encounter the messaging, such as billboards and bus shelter placements. These communication efforts were led and coordinated by the City's Joint Information Center, which housed the public information officers from all City agencies. Working

FIGURE 12.2 Early pandemic risk communication campaigns focused primarily on protecting the families and loved ones of city residents from disease transmission.

across agencies ultimately helped to amplify the critically important messages across the city.

Communication efforts for vaccination campaigns

The BCHD's communications plan for the rollout of the COVID-19 vaccines considered a number of factors but predominantly focused on vaccine scarcity and vaccine hesitancy. Still, beyond what these plans considered, perhaps most critical was the approach it endorsed—a grassroots approach utilizing trusted messengers within the community. Several different, yet interlocking, communications campaigns were developed. These were organized using a framework informed by an individual's eligibility to be vaccinated based on the priority matrix developed by the MDH, as well as an individual's personal desire to be vaccinated. Internally, these communications campaigns were organized into four key groups—Where, When, Why, and Wait—with a quote suggesting their attitude toward the vaccine (Box 12.1).

Due to vaccine scarcity, having an appointment to secure a COVID-19 vaccine was necessary, and the communications of the BCHD reflected the need to sign up for vaccination. Initially, shifting guidelines related to the priority matrix and high public demand meant that much of the "Where" and "When" populations were addressed via social media including Twitter, Facebook, Reddit, and Instagram, and traditional media outlets such as press releases and press conferences, where updates related to the number of vaccines available at any one location could be given in the same day. Promotion of the vaccine priority matrix—including those who were at higher risk of developing serious COVID-19-related illness and therefore were eligible to be vaccinated earlier than others—was of the utmost importance. Calendars indicating where vaccination clinics were located, their hours of operation, who was eligible, and any updates in vaccine supply, were developed and disseminated via social media on a daily basis (Fig. 12.3). A call

---

BOX 12.1 Details and suggested attitudes of the four key groups targeted by vaccination campaigns, as conceptualized by the BCHD.

Where?	"Where can I get the vaccine?" *I am excited to receive the vaccine, and I am currently eligible to receive it based on the priority matrix.*
When?	"When can I get the vaccine?" *I am excited to receive the vaccine but am not currently eligible to receive it based on the priority matrix.*
Why?	"Why should I get the vaccine?" *At some point, I will be eligible to receive a vaccine, but I'm not convinced that I need it, am skeptical of the benefits, or have concerns about the safety of the vaccine.*
Wait.	*This population was not interested in being among the first groups to be vaccinated and regardless of communications strategy, would adopt a "wait-and-see" approach before seeking vaccinations.*

FIGURE 12.3 Communication materials for COVID-19 vaccination campaigns developed by the Baltimore City Health Department.

center staffed by BCHD employees was also set up to answer questions regarding eligibility and the availability of vaccines.

Months before the first vaccine was available, the BCHD team members knew it would be crucial to begin addressing vaccine hesitation, especially in communities of color. These efforts were informed, in part, by previous departmental efforts to promote the uptake of seasonal influenza vaccines. In the fall of 2020, the BCHD held a number of virtual focus groups with Baltimore City residents to design and develop a flu vaccination campaign. Outputs from these efforts later informed what eventually became the BCHD's COVID-19 vaccine communications campaign. For instance, focus grouping with residents in the communities most impacted by seasonal influenza overwhelming suggested that the key to effective communication was to communicate "authentically." These conversations inspired the BCHD to develop the "It's Baltimore Versus COVID" campaign—a series of digital and print media campaigns, including a local television commercial, featuring well-known residents from a variety of backgrounds including local government leaders, religious leaders, rap artists, youth ambassadors, and community association presidents (Fig. 12.4). This campaign purposefully adopted an "us-versus-them" sports analogy to engender feelings of togetherness and civic pride. Importantly, rather than taking a "one-size-fits-all" messaging approach, the Baltimore Versus COVID campaign featured a variety of harm reduction messages—often using familiar language, idioms and directly quoting featured subjects—to authentically resonate with different populations residing in the city. Messages included in this campaign highlighted the importance of getting vaccinated, wearing a mask around others, frequent hand washing, and getting tested for COVID-19, while emphasizing a communal approach to successfully navigating the pandemic (e.g., "We're all in this together").

Over time, as vaccines became more available and eligibility restrictions began to relax, communications began to shift away from the concerns of the "Where" and "When" populations, and more towards encouraging vaccinations in the "Why" and "Wait" communities. Thanks to community feedback solicited via focus groups and previous public health campaigns, the BCHD knew that to address vaccine hesitation, in-person communication from trusted messengers would be vital. In addition to BCHD staff collectively participating in over 150 virtual town halls, where staff could answer questions from the community regarding the vaccines, the BCHD launched a vaccine ambassador program that trained and compensated engaged members

FIGURE 12.4 Communication materials from the "It's Baltimore Versus COVID" communication campaign developed by the Baltimore City Health Department.

of the most impacted communities to canvas their own neighborhoods and answers questions about the COVID-19 vaccines. Vaccine ambassadors would deploy in the neighborhood surrounding a pop-up community vaccination clinic a few days before the event to provide notification of the vaccination clinic and to answer questions and concerns residents had about vaccination. Print materials were developed for the vaccine ambassadors and shared with the health care provider community, as well.

Communications efforts addressing mis- and disinformation

The BCHD understood, early on, its own limitations in convincing a skeptical public to seek vaccination. These limitations were due to, among other factors, a historic mistrust of the medical community, a distrust of the government, or both. Digital communications during the pandemic response sought to support the on-the-ground work of the vaccine ambassador program by addressing and pushing back against online misinformation and disinformation campaigns related to the COVID-19 vaccines. Frequently repeated community misapprehensions about the vaccines overheard by vaccine ambassadors during their outreach were relayed to BCHD leadership. This direct feedback was then incorporated into mass communication campaigns on social media, putting into motion a rapid-response feedback loop to address concerns head-on. During this time, BCHD social media channels employed a number of "meme" campaigns, using campy Shutterstock images, authentic and relatable situations, and simplified harm reduction messaging, to address these concerns in a lighthearted manner (Fig. 12.5). The meme campaign's goal was to specifically address problematic rumors and behaviors related to COVID-19 and the COVID-19 vaccines, as well as to introduce those messages in a format that could be shared among friend

FIGURE 12.5 Materials from the COVID-19 meme campaign to combat misinformation, developed by the Baltimore City Health Department.

groups, specifically in text message conversations and other "offline" avenues. Acknowledging the hostile online environment of public health messaging at the time, the offline strategy was identified as one method with fewer barriers to overcome because in this scenario a close friend, likely a trusted messenger, would be the promoter of the harm reduction message instead of a traditional public health messenger. This strategy was employed to great effect in Baltimore City, which among other positive outcomes led to a dramatic increase in the following of the BCHD online, as well as the publication of several positive local, national, and international news stories related to these efforts. To address the concerns of the "Wait" population, the BCHD primarily relied upon information from its public COVID-19 dashboard and the work of its epidemiologists, which showed that Blacks between the ages of 25–50 were lagging in their vaccination rate. After seeing greater vaccination rates of its older adult population, a campaign was developed to target young parents, who statistically were likely to know at least one older adult that had been vaccinated.

Conclusion: lessons learned for effective urban public health campaign

Baltimore fared better than other, comparable jurisdictions in terms of COVID-19 incidence, mortality, case fatality, and vaccination rates for residents over 12 years of age and the varied forms of messaging were key to the city's success [6]. Given this success, there are many key takeaways surrounding how to best effectively message to urban populations in times of public health emergency. First, all messaging should be clear and transparent. Health authorities must share that there are unknowns but also reassure the public that when more is known they will be updated and informed.

Second, authorities should have plans that use a variety of communication mediums. In oral communications, authorities should strive to be empathetic and relatable. To this end, stressing that "we are in this together" and "we want to support residents as we all make hard decisions to protect ourselves and our families amidst the uncertainties" can be effective and impactful. In written communications, authorities should have a strategy and work plan—especially as these allow for process and outcomes evaluations—but these should allow for some degree of flexibility as more information becomes known.

Third is that tailoring the message to the population is equally important to the message itself. To do this, authorities must know the population they are serving. Utilizing focus groups can be immensely helpful for better understanding the population and for refining both messages and message delivery—as the target populations are the experts on what messages will resonate, not health authorities. Authorities must acknowledge the concerns of a population, identify their hopes for their families and quality of life, determine their daily struggles, and incorporate their feedback into messaging. This also means meeting the community where they are. In Baltimore, this meant recognizing that there are valid and deep-seated reasons for mistrust in government and health care. While the BCHD was not able to right these past wrongs, it did work to create a new path forward forged in acknowledgment and truth.

Additionally, authorities should always identify the best message delivery mechanism for the community that they are trying to reach. Slicing the same message in different ways and different formats is critical to reaching a diverse audience. Relatedly, some populations will respond better to different messengers and authorities should utilize messengers that the specific, target population will relate to. So, while the messenger may change, the message remains the same. And, in all this, authorities should not forget to also have fun with the messaging. Doing so may garner interest in unexpected ways and spark interesting discussions and engagement that otherwise would not have happened.

Finally, and most importantly, authorities should spend time serving in the communities they are hoping to reach, using these opportunities to develop trust and show a genuine commitment to serving the community that benefits from the public health messaging.

References

[1] Samuels B. Segregation and public housing development in Cherry Hill and Westport: historical background. Annapolis: Maryland State Commission on Environmental Justice and Sustainable Communities 2008.

[2] Williams D.R. Race, racism, and Baltimore's future: a focus on structural and institutional racism. Baltimore: Johns Hopkins Urban Health Institute; 2016.

[3] Williams DR, Collins C. Racial residential segregation: a fundamental cause of racial disparities in health. Public Health Reports 2001;116(5):404−16. Available from: https://doi.org/10.1093/phr/116.5.404.

[4] Federal Emergency Management Agency. IS-0100.C: an introduction to the incident command system. ICS 100. Washington, DC: Department of Homeland Security; 2018.

[5] UnitedStatesZipCodes.org. Cities in ZIP Code 21215, <https://www.unitedstateszipcodes.org/21215/>; 2022 [accessed 12.08.22].

[6] Lee K.H., Marx M. Baltimore City: county-level comparisons of COVID-29 cases and deaths. Baltimore: Baltimore City Health Department; 2021.

Conclusion

Matthew R. Boyce and Rebecca Katz

We have seen a growing recognition of the importance of involving subnational actors in health security and pandemic preparedness and response efforts over the past several years. The World Health Organization (WHO) held a landmark consultation in October 2008 on the role of cities in addressing public health crises and implementing the International Health Regulations (IHR 2005) [1]. However, this was followed by a period of stagnation before the WHO held a subsequent meeting nearly a decade later, and before specific guidance was developed for use by local authorities during and after the COVID-19 response [2–4]. This subnational guidance was notable for several reasons. At the most basic level, it was a deviation from the norms and mandate of the WHO, which is an organization composed of member states (i.e., nations and territories), not subnational actors, such as local authorities. But beyond this, and perhaps more importantly, it was a clear acknowledgment that local authorities have a role—and a critically important one at that—in achieving, local, national, and global health security.

While these are important normative developments that could signal strategic shifts in health security efforts, they come from a higher level of governance that is inherently less involved in the actual implementation of health security and response activities. This is where the value of this book lies. Many chapters give a voice to the individuals who coordinated and implemented the response efforts to this historic public health emergency.

As readers have no doubt realized by now, no two cities are the same. Each operates in its own unique legal, political, economic, social, and epidemiologic contexts. Indeed, this is why localized approaches are so important. Readers, however, are also likely to have realized that local authorities often used many of the same broad approaches to respond to the pandemic—adapting them to their context as necessary and appropriate. This raises important questions of whether there is a suite of themes and best practices that can be identified.

Across many, if not all the chapters, there are two prevalent themes: the repurposing of existing capacities and infrastructure, and the use of vaccines.

Many cities repurposed existing capacities and infrastructure to support the pandemic response. These efforts included the use of physical urban infrastructure—such as the use of stadiums, hotels, and schools to support diagnostic testing, vaccination, and quarantine and isolation efforts—as well as other resources, such as health-care workforces. A compelling question, of course, is whether this has always been a good thing or whether an indication of insufficient resources and infrastructure to begin with. For instance, many public health workforces were repurposed to support the pandemic response at the detriment of other important programs and services, such as those supporting maternal and child health, sexually transmitted disease control, and environmental health. What seems likely to be important moving forward will be considering how to plan for and build this resilience-promoting flexibility into urban infrastructure. Indeed, in our increasingly urban, globalized world—and one that faces an increasing threat from emerging infectious diseases [5]—there will certainly be epidemics and pandemics in the future. How well cities are able to absorb and respond to these events will require deliberate and intentional planning, and coordination across a wide range of sectors and disciplines.

Another ubiquitous theme across chapters was the importance of vaccines and vaccination campaigns. The speed with which vaccines that protected against COVID-19 were developed is nothing short of astonishing. When the disease was first detected in late 2019, few could have predicted that vaccines would have been developed by the end of 2020. Indeed, before COVID-19, the fastest that a vaccine had previously been developed was for mumps in the 1960s—an effort that took 4 years from viral sampling to vaccine approval [6]. But while the development of the COVID-19 vaccines is a testament to the power of science and modern medicine, vaccines are not valuable unless they are given to people and used to inoculate them against disease. Put more simply, it was the vaccinations and not the vaccines themselves that made a difference in controlling the pandemic. And, as referenced in many of the chapters, these efforts required considerable amounts of preparation and resources. From developing plans and strategies for the rational and ethical rollout of vaccines, to identifying venues and training workforces to support mass vaccination campaigns, to considering the logistics required to safely store the vaccines and maintain the cold chain, to monitoring the distribution of vaccines, to intentionally reaching out to subpopulations to promote vaccine equity, there are many considerations that demand attention for successfully turning vaccines into vaccinations. And capitalizing on their deep knowledge of the resources and populations in their cities, local authorities were uniquely positioned to ensure that these efforts were—and will continue to be—effective, efficient, and ultimately successful.

There are other themes that also appear in many of the chapters included in this book. For instance, the chapters on Kawasaki City (Chapter 1), Lagos (Chapter 2), Singapore (Chapter 5), and Las Vegas (Chapter 7) all reference the importance of robust legal preparedness. As seen with Lagos and Singapore, this type of preparedness can be informed by preceding responses and is essential for promoting a coordinated, efficient, and rapid response. It can also support the swift promulgation of emergency legislation to support response efforts, as was the case in Las Vegas when certain workforce licensing requirements were relaxed to surge the workforce.

The value of technology and information systems in supporting outbreak response is another common theme among the included chapters. As alluded to in the chapters on Lagos (Chapter 2), Idlib (Chapter 4), Singapore (Chapter 5), Quezon City (Chapter 6), Dhaka (Chapter 8), and Cape Town (Chapter 10), telemedicine and digital health solutions played an important role in many urban pandemic responses and helped to ameliorate the stress on health-care systems and capacities. Further, the development and use of mobile applications supported a range of activities from contact tracing, to vaccine registration, to compensating workforces.

Relatedly, many chapters, including those focusing on the responses in Beirut (Chapter 3), Quezon City (Chapter 6), and Dhaka (Chapter 8), discussed how automated and interoperable information systems supported the pandemic response. When systems are able to communicate with one another, they can support many aspects of pandemic response, including genomic surveillance, health system capacity (e.g., the availability of hospital beds), access to electronic medical records, and monitoring adverse medical events. We imagine that ongoing technological advancements will continue to allow for innovations in urban health security.

Community engagement and mobilization was another frequently discussed theme in several chapters. Whether it was the support provided by lay-persons in Kawasaki City (Chapter 1), volunteers in Idlib (Chapter 4), community health workers in Dhaka (Chapter 8) and Cape Town (Chapter 10), or "COVID Warriors" in Dharavi Slum of Mumbai (Chapter 11), working closely with communities is a clear best practice for successfully implementing pandemic response activities. And, as discussed in the chapter on Baltimore (Chapter 12), it is essential that authorities not only engage with communities during emergencies but spend time working with and serving them beforehand to demonstrate their genuine commitment to those people and develop trust.

The experiences from Lagos (Chapter 2), Idlib (Chapter 4), Singapore (Chapter 5), Quezon City (Chapter 6), Dhaka (Chapter 8), New Orleans (Chapter 9), and Baltimore (Chapter 12) also underscore the importance of risk communication. Whether this pertains to the internal communication efforts between various departments, agencies, and authorities, or public-facing external communication efforts, good risk communication holds the potential to improve coordination between relevant authorities, instill a sense of confidence and trust in the public, promote the uptake and compliance with outbreak response interventions, and address the threats of mis- and disinformation.

Several chapters also discussed the importance of "hyperlocal" outbreak response efforts for ensuring that important information and services reach vulnerable populations. While the causes of vulnerabilities can vary—ranging from biological factors, to socioeconomic considerations, to historic legacies—the experiences from Quezon City (Chapter 6), New Orleans (Chapter 9), Cape Town (Chapter 10), Mumbai (Chapter 11), and Baltimore (Chapter 12) suggest that these efforts work best when they are culturally sensitive, equity-focused, and proactively engage these populations instead of expecting them to overcome significant access barriers.

One final theme discussed in several chapters was that dynamic and unstable contexts do not necessarily preclude successful outbreak response. Authorities in Beirut were required to respond to not one, but two public health emergencies at the same time (Chapter 3). Readers have learned about how authorities in Idlib conceptualized and implemented a pandemic response in a conflict zone (Chapter 4). And the experiences from Cape Town underscore how residual effects from pandemic response activities can cause social unrest (Chapter 10). Still, despite these challenges, these cities were able to mount effective COVID-19 response efforts.

In addition to these themes and best practices identified across chapters, there are several additional themes that must be acknowledged to ensure that cities around the world are better prepared for future infectious disease outbreaks. The first is recognizing the shared infectious disease risk between cities. While the specific jurisdictional responsibilities and authorities of local governments vary considerably in cities around the world, what remains constant is cities act as the conduit for the spread of disease in our globalized world. Indeed, we witnessed this throughout the COVID-19 pandemic whereby the disease spread around the world through global cities, and around countries through cities of national economic and cultural significance. Recognizing that disease will flow between cities in this way, one practical way to ensure a faster and more efficient response is to develop new surveillance systems and information-sharing programs as a means of promoting a rapid and coordinated response. For instance, aircraft wastewater surveillance, already being piloted, could prove a useful tool for surveilling and monitoring the global spread of certain pathogens [7].

It is also growing increasingly clear that those working in public health, policy, and security may need to reconsider what aspects are measured by preparedness and response capacity assessments, and the granularity at which those assessments are conducted. At the time of writing, a majority of the emerging evidence suggests preparedness assessments were not well attuned to actual performance during the pandemic response [8–11]. That is, higher preparedness scores did not translate to better responses or health outcomes during the COVID-19 pandemic. Further, before the COVID-19 pandemic, leading health security frameworks operated primarily at the national scale—assessing broad levels. While this scope may be appropriate for certain contexts, it is unsuitable for others.

Take, for instance, the US Joint External Evaluation of the IHR Core Capacities, which found the country had reasonably developed risk communication capacities [12]. In a large, diverse, federalist country like the United States, what does this result mean in reality? Is it reasonable to infer that this capacity is uniform across the country, or does it vary across different environments? This is particularly important in federal countries that allocate various health authorities and responsibilities to subnational levels of government, as the capacity and performance of subnational governments can vary considerably. While incremental changes have already been made to the JEE to be more inclusive of subnational entities [13], global health officials may need to further reconsider and potentially reconceptualize what is measured by these assessments, and at what scale, to ensure health security at all population levels. Doing so could help to ensure that our cities and world are robust and resilient to everyday health threats, as well as those that will strain and challenge capacities regardless of preparedness levels.

Relatedly, much as it will be important to reconsider the level of assessments, it may also be prudent to reconsider the level of governance in outbreak response. Health security is currently governed primarily at the global and national levels. But cities and local authorities, with increasing frequency, are taking the lead in addressing other transnational issues, such as environmental issues and climate change [14]. This naturally raises questions about whether cities can assume a more prominent role in the governance of health security threats as well. One practical way to do so could be the creation of collaborative city networks that would promote collaboration between local authorities and the sharing of technical expertise to promote public health preparedness and response. To note, such actions would be well aligned with Article 44 of the IHR, which outlines how countries are expected to collaborate with one another to implement the provisions of the legally binding framework [15]. Several mayoral networks also assumed this role, at least temporarily, during the COVID-19 pandemic [16]. But many of these networks existed before the pandemic—with their own mandates and agendas—and have since pivoted back to their original scopes, leaving a void. Additionally, new forms of urban networks, comprising public health authorities, as opposed to mayors, could also be developed. Such a network, supported through MOUs, could facilitate the informal sharing of data, lessons, and alerts, especially between highly connected, global cities. And, while such a proposition may have seemed far-fetched at one time, recent trends in governance and diplomacy suggest that this could be a viable policy option moving forward. For example, in 2022, the US Department of State named its first Special Representative for Subnational Diplomacy—a position that would seek to engage local partners, foster connections among cities, develop solutions and partnerships to address key issues, and strengthen diplomatic ties to cities [17].

Last, while it is difficult to make assertions regarding the relative importance of certain health security capacities compared to others, experiences from the COVID-19 pandemic have made it clear that the public health workforce and associated human resources for health are among the most critical. Without a robust and resilient public health and health-care workforce, cities stand little chance of controlling an outbreak, as these individuals are integral to all aspects and activities of epidemic and pandemic response. Still, the COVID-19 pandemic has resulted in substantial amounts of emotional, mental, and physical strain on these individuals, and many expect considerable amounts of attrition in the coming years. For example, a study conducted by the US Centers for Disease Control and Prevention and published in 2022 found that 44% of public health workers in the country were considering leaving their jobs within the next five years [18]. Because of this, there is a compelling need for long-term strategies to develop the public health workforce in cities around the world. Importantly, these strategies must be accompanied with dedicated funding to ensure these efforts to improve preparedness are not performative or short-lived, but practical and sustainable.

It seems appropriate to close this volume with grateful acknowledgment of all of the efforts of local and subnational authorities throughout the pandemic. These individuals worked tirelessly throughout the pandemic, often under trying circumstances, to protect the public's health. They also represent an invaluable wealth of knowledge. It is not enough to simply identify lessons from their experiences in responding to the COVID-19 pandemic, but these lessons must also be learned as a means of preparing our cities and world for future outbreaks. And, just as Singapore learned from the first SARS epidemic, Lagos from the Ebola epidemic, and New Orleans from the Zika epidemic, it seems reasonable to assume that cities represent a viable environment for this learning to occur and to make pragmatic change.

References

[1] World Health Organization. Cities and public health crises: report of the international consultation. WHO/HSE/IHR/LYON/2009.5. Lyon: World Health Organization; 2009.

[2] World Health Organization. High-level conference report: preparedness for public health emergencies challenges and opportunities in urban areas. WHO/WHE/CPI/LYO/2018.64. Lyon: World Health Organization; 2019.

[3] World Health Organization. Practical actions in cities to strengthen preparedness for the COVID-19 pandemic and interim checklist for local authorities. Geneva: World Health Organization; 2020.

[4] World Health Organization. Framework for strengthening health emergency preparedness in cities and urban settings. Geneva: WHO; 2021.

[5] Jones KE, Patel NG, Levy MA, Storeygard A, Balk D, Gittleman JL, et al. Global trends in emerging infectious diseases. Nature. 2008;451(7181):990−3. Available from: https://doi.org/10.1038/nature06536.

[6] Ball P. What the lightning-fast quest for COVID vaccines means for other diseases. Nature. 2021;589(7840):16−18. Available from: https://doi.org/10.1038/d41586-020-03626-1.

[7] Jones DL, Rhymes JM, Wade MJ, Kevill JL, Malham SK, Grimsley JMS, et al. Suitability of aircraft wastewater for pathogen detection and public health surveillance. The Science of the Total Environment. 2023;856:159162. Available from: https://doi.org/10.1016/j.scitotenv.2022.159162.

[8] Haider N, Yavlinsky A, Chang Y-M, Hasan MN, Benfield C, Osmanet AY, et al. The Global Health Security Index and Joint External Evaluation score for health preparedness are not correlated with countries' COVID-19 detection response time and mortality outcome. Epidemiology and Infection. 2020;148(e210):1−8. Available from: https://doi.org/10.1017/S0950268820002046.

[9] Abbey EJ, Khalifa BAA, Oduwole MO, Ayeh SK, Nudotor RD, Salia EL, et al. The Global Health Security Index is not predictive of coronavirus pandemic responses among Organization for Economic Cooperation and Development countries. PLoS One. 2020;15 (10):e0239398. Available from: https://doi.org/10.1371/journal.pone.0239398.

[10] Baum F, Freeman T, Musolino C, Abramovitz M, De Ceukelaire W, Flavel J, et al. Explaining COVID-19 performance: what factors might predict national responses? British Medical Journal. 2021;372:n91. Available from: https://doi.org/10.1136/bmj.n91.

[11] Khalifa BA, Abbey EJ, Ayeh SK, Yusuf HE, Nudotor RD, Osuji N, et al. The Global Health Security Index is not predictive of vaccine rollout responses among OECD countries. International Journal of Infectious Diseases. 2021;113:7−11. Available from: https://doi.org/10.1016/j.ijid.2021.09.034.

[12] World Health Organization. Joint External Evaluation of IHR Core Capacities of the United States of America: mission report, June 2016. Geneva: World Health Organization; 2017.

[13] World Health Organization. Joint External Evaluation Tool: International Health Regulations (2005). 3rd ed. Geneva: World Health Organization; 2022.

[14] Acuto M. City leadership in global governance. Global Governance. 2013;19(3):481−98.

[15] World Health Organization. International Health Regulations (2005). 2nd ed. Geneva: World Health Organization; 2005.

[16] Boyce MR, Katz R. COVID-19 and the proliferation of urban networks for health security. Health Policy and Planning 2021;36(3):357−9. Available from: https://doi.org/10.1093/heapol/czaa194.

[17] Blinken AJ. Naming Ambassador Nina Hachigian as Special Representative for Subnational Diplomacy. US Department of State: Washington DC; 2022.

[18] Bork RH, Robins M, Schaffer K, Leider JP, Castrucci BC. Workplace perceptions and experiences related to COVID-19 response efforts among public health workers − public health workforce interests and needs survey, United States, September 2021 − January 2022. MMWR. Morbidity and Mortality Weekly Report. 2022;71:920−4. Available from: https://doi.org/10.15585/mmwr.mm7129a3.

Index

Note: Page numbers followed by "*f*" and "*t*" refer to figures and tables, respectively.